To

LEWIS MUMFORD

whose great contribution

to planning thought

was the main inspiration

which prompted this book

First published in mcmlxv
by Faber and Faber Limited
24 Russell Square London W.C.1
New and revised edition mcmlxviii
Printed in Great Britain by
Western Printing Services Limited Bristol

S.B.N. 571 04014 4

JOHN TETLOW & ANTHONY GOSS

Homes, Towns and Traffic

WITHDRAWN
FROM STOCK

FABER AND FABER

24 Russell Square

London

Homes, Towns and Traffic

Contents

Illustrations

PLATES

11

ILLUSTRATIONS

ILLUSTRATIONS

FIGURES IN TEXT

13

ILLUSTRATIONS

14

ILLUSTRATIONS

ILLUSTRATIONS

Acknowledgements

Acknowledgements are made to the many public authorities and private firms of architects and town planners who have so willingly given information, drawings and photographs. The sources of line drawings have, where appropriate, been acknowledged in the text. In most cases, line drawings have been redrawn from the originals to make them more suitable for publication here. Figure 8 is based on drawings from *Town Planning and Road Traffic* by permission of the publishers, Edward Arnold, and Figure 50 is reproduced by permission of *Official Architecture and Planning*.

The sources of photographs used as plates, and the photographers, where known, are credited separately on pages 267 and 268.

We are deeply indebted to our secretaries—in particular to Rosemary Thornhill and May Metcalfe—for the unfailing patience and skill with which they have transcribed our manuscript.

Preface

Since we prepared the first edition of *Homes, Towns and Traffic* over four years ago there have been many significant changes in the techniques of planning, designing and building towns and in the degree of attention focused on traffic and transportation. Our intention in this new edition is as far as possible to take account of these developments: indeed, as might be expected, some of our own ideas and opinions have changed. We have also received comments from many town planners, architects and others whose work was referred to, and reviewers have suggested the inclusion of other aspects of the subject in later editions. For these we offer our sincere thanks: if we have not adopted all the suggestions made they have nevertheless helped us materially. Substantial revisions have in fact been made from Chapter 4 onwards, particularly in the examples discussed and illustrated: Chapters 5 to 8 have been completely rewritten.

Whilst the superstructure has been modified the foundations remain: the purpose of the book is unchanged. It is to examine how far town planning has gone towards its true objectives, which in our view are the welfare of man, his health, happiness and convenience. It could also be said, with considerable truth, that the objectives of planning are to achieve maximum 'efficiency' in the strictly mechanistic sense. We detect a certain conflict between these views, a conflict which is reflected in controversy among town planners and associated professions as to what town planning is all about and the delimitation of the sphere of operations of those engaged in it.

No-one should doubt the need for a more scientific basis and a greater technical competence in reshaping our habitat. Recent technological advances must be more effectively applied to and reflected in the environment in which we live, work, and take our ease. We suggest that the basic problem to be overcome within the second half of the twentieth century is so to fashion

19

our environment that man is served and not dominated by scientific and technical achievements. This has been the broad, progressive yet humanitarian message of the great pioneers of modern town planning since the beginning of the century–men like Patrick Geddes, Raymond Unwin, Le Corbusier, Patrick Abercrombie and Lewis Mumford. It is already obvious that there is a serious danger that as the complexities of town planning become more fully recognized, progressive and humanitarian ideals become submerged in the very urgency of traffic and housing problems, lost in a welter of economic, administrative and legal technicalities. For the technocrat the phrase 'planning is for people' may produce a sneer. It may be trite: it can be twisted to mean almost anything. But its simple truth is unassailable.

This is the approach which has shaped our book. We have tried to outline the background, the pioneering ideas, the problems, experience and lessons of designing and building towns, concentrating on Britain as the main area for case study. No attempt has been made to cover comprehensively theory and practice throughout the world but examples from other countries have been used by way of additional illustration. Our account of the planning process is directed in particular towards the interested layman and the university student seeking an introduction to the subject. We have therefore tried to avoid jargon and to present problems in everyday language but without shrinking from going into technical detail where necessary.

We have generally placed emphasis on what has actually been done rather than what is 'in the pipeline' because what exists can be evaluated with varying degrees of sophistication. The individual citizen can go and judge for himself: deeper studies can confirm or confound the assumptions of the designer. This is the test–are the activities of so-called civilized men producing the best environment attainable within the resources of modern society and in step with achievements in other fields? Are even tolerably acceptable results being attained? Only the most complacent and insensitive would at present affirm that they are. There is still a great deal to be learnt if we want to do better.

CHAPTER 1

The Transport Revolution

The art of living together is the hallmark of civilization. Throughout history, men have grouped themselves together for safety, for the exchange of services, food and goods, for worship, and for social intercourse. The town is the highest expression of these mutual needs; it became the seat of administration and justice, the market and the focus of manufacture, a nursery of the arts, religion and science—in every way a utility for collective living. Whilst towns remained small this was clearly reflected; there was face-to-face association of a limited number of people, relatively little movement between various parts of the country and co-operation in manufacture by simple craft methods. Communal life centred, quite naturally and spontaneously, on the parish church, the market and the guildhall. And parts, at any rate, of our earlier towns displayed the pride of their citizens in making buildings beautiful as well as useful. The houses in the medieval High Street vied with each other in their size and in the refinement and decoration of their craftsmanship; the community expressed its corporate spirit in its churches, town hall and almshouses.

But individual and collective needs and ambitions change; a major problem today is to find a way of communal life which reconciles our reliance on motor vehicles for personal, public, industrial and commercial use with the need for safety, privacy and quiet in our homes. The purpose of this book is to consider how this can be done. It is first necessary to go back and trace briefly how the present situation came about. Comprehensive accounts of the historical development of towns can be found elsewhere, particularly in the works of Lewis Mumford, Thomas Sharp and the late Professor Sir Patrick Abercrombie; we shall only give a sketch of what has happened in the last 200 years in Britain, with particular reference to changing methods of transport.

The English town, as a place to live in, was perhaps at its best in the seventeenth and eighteenth centuries. By a happy chance the country's rise to commercial greatness, largely founded on the production of wool and woven cloth, coincided with the full development of the Renaissance ideas of order and seemliness; the expansion and rebuilding which took place everywhere was infused with a spirit of urbanity. Terraces, streets, squares were built, not always to a unified design but in the same style and of much the same height. Individual ostentation and eccentricities were not expected or sought for; a great degree of uniformity was acceptable. It was understood that the amenities of the group as a whole contributed to those of each house.

The grandiose schemes imposed on European towns by absolute monarchs and military dictators never appeared here; instead, our tradition was formed on the more modest scale of domestic grace and elegance. The change, started in the quadrangles of Oxford and Cambridge and the Piazza of Covent Garden, spread through London and to the resorts of Bath, Cheltenham, Tenby, Brighton and a score of others; it attained one of its greatest glories in the New Town of Edinburgh in 1767, and produced splendid streets and squares not only in Liverpool, Newcastle, Plymouth and Bristol but in many of the smaller provincial towns.

Little of this was done under civic control. Apart from the Edinburgh New Town, the only large scheme carried out with civic authority was the Regent Street and Regents Park development in London. The earlier squares were built, a few houses at a time, for individual clients. Then came the building of whole quarters on the initiative of landowners and speculative builders, who, whilst they amassed considerable fortunes in the process, were content to subscribe to the good manners of the day in subordinating personal idiosyncrasies to the principles of order and good design. They were not, it is clear, attempting to achieve immortality by building masterpieces of architecture but quite soberly providing houses to live in as comfortably and pleasantly as possible according to the standards of the time.

The pattern was not universal; these were the houses, the terraces and the squares for the gentry and the rich merchants. But behind them or near at hand more modest streets were built on similar lines for the shopkeepers and small tradesmen. The older houses deserted by the upper classes for the elegant new ones were taken over by the growing numbers of wage earners who came into the towns; in alleys and courts behind the better streets the 'lower orders' congregated in the worst conditions of squalor and decay. Even the best of the new buildings were deficient in sanitation; water supply was indifferent and the disposal of sewage and refuse primitive; the cleaning and repair of public streets was quite unorganized as late as 1761 and improved only slowly. But though the English town was no earthly paradise it was a remarkably civilized place to live in. The best of the residential districts provided–and some of them still provide–the finest and most gracious environment to be found anywhere in the world; the skilled man generally had a tolerable house or a part of house. The towns, though growing, were still small, and fields, orchards and commons were near at hand.

The change came as the result of the rapid growth of the factory system. The first stage of the Industrial Revolution took place with water wheels as the source of power; by 1831 cotton textiles were established as the main manufacturing industry employing 800,000 people. Between 1760 and 1830 the population doubled, from 8 million to 16 million; Birmingham, Sheffield and the cotton towns grew more rapidly between 1821 and 1831 than ever before or since. Many other towns grew and villages became towns. With the introduction of steam power the change became truly revolutionary; Britain quickly became the leading industrial country in the world and population and employment concentrated around the coal and iron deposits, principally in the North of England and Midlands. In 1700 the counties with the highest populations were Somerset, Wiltshire, Gloucestershire, Northamptonshire and Middlesex (which included

London). By 1800 only Middlesex remained in this list, the places of the others had been taken by Lancashire, the West Riding, Staffordshire and Warwickshire.

The effect on living conditions in towns was catastrophic; the overwhelming growth of the working class population brought disaster. Houses were needed for them and houses were built—the greatest possible number of the smallest type on the smallest area of land, packed tightly round the factories and the mines, with no open space, no sanitation and the most primitive services. In Liverpool, one family in every five lived in a cellar. In these conditions the town as a community—an expression of civilized life—broke down. By the middle of the nineteenth century: 'the idea of a town as a focus of civilization, a centre where the emancipating and enlightening influences of the time can act with rapidity and with effect, the school of the social arts, the nursery of social enterprise, the witness of the beauty and order and freedom that men can bring into their lives, had vanished from all minds.'[1] The town became instead a necessary evil—a workplace and a barracks for the 'hands', where women and children, as well as men, toiled like animals for incredibly long hours for subsistence wages, and from which those who could escaped at the end of the day in search of purer air, light, space and quiet. The picture was not wholly dark. Even in these savage conditions communal spirit struggled to the surface and flowered bravely in the life of Church and Chapel, the beginnings of universal education, the trades unions, mechanics institutes, friendly societies and the co-operative movement.

THE TRANSPORT REVOLUTION

At the same time a revolution in transport and communication had been going on; its effects have been insufficiently recognized and in the context of this book, call for special emphasis.

In the early 1800's the main roads of England, the roads between towns, between London and the ports, were the finest in Europe. Notwithstanding the strange system of turnpike trusts

[1] J. L. and B. Hammond, *The Town Labourer*.

which, at their worst, could be remarkably inefficient and corrupt, the sound methods of Telford and Macadam had founded a network of splendid roads over the greater part of the land on which the mail coach services ran swiftly and smoothly. The minor routes fell far below this high standard; they were left to the parish, the amateur, unpaid parish surveyor, the statute labourer and the pauper road mender. Within towns, traffic was relatively small. Hackney coaches had been permitted in London from 1634 and in the same year sedan chairs came into use. In 1837, 600 stage coaches were licensed to ply between London and towns within nineteen miles of it, and in 1829 Shillibeer started his first omnibus from Paddington Green to the Bank. But by today's standards there was little wheeled traffic, for distances between parts of the town were not too great to walk.

For the carrying of goods, the canal system was virtually complete by 1830. Starting in South Lancashire, waterways had been carried across the Pennines to link up with the Humber. They had spread through the Midlands and extended to join the Severn and the Thames. They served the already immense coal trade but even this was subordinate to the movement of general merchandise, raw materials to the manufacturing districts, finished goods, food and building materials to London and the ports.

The picture was changed radically by the coming of the railways. Dating from Stephenson's triumph at the Rainhill locomotive competition in 1829 and the introduction of the wrought-iron edge rail, a wild mania for railway building swept the country; vast numbers of lines were proposed which were never built. But, by 1848, 5,000 miles of track had been laid and were in use in the United Kingdom, the greater part of the whole railway system of modern Britain. The effects of this on towns and on urban living were numerous and profound. Towns which became railway centres flourished; Swindon, hitherto a small market town, and Crewe, a hamlet with less than 500 inhabitants, are examples. Locomotive workshops were established, railway clothing and printing factories were added, the essential services of water and gas were provided by the railway company. Since many of these towns were also junctions, houses had also to be built for operating staff. This growth and dependence on the railway went on up to the outbreak of the First World War. By contrast, the towns which failed to attract the important lines, or thrust them away, stagnated into unimportance. But all towns through which railways ran were affected to a greater or lesser degree. Communities were divided and segregated, as never before, by the steel barriers of the permanent way, its embankments and cuttings; the smoke and noise of trains was added to those of factories and workshops in the centres of towns, and penetrated into the heart of the countryside.

An even more fundamental change was that Britain became a mobile nation. Men and women could now travel as never before to work or on pleasure. As early as the eighteenth century the rich merchants had begun to move out south and west to the suburbs; in the 1820's there was wholesale migration from London of 'City men', who sometimes lived as far afield as Brighton; this now became a steady stream. The dormitory suburbs grew on all sides; Bournemouth, Southport, Harrogate, Leamington and their counterparts developed as towns for retired people. This migration was not for the common herd, whose houses remained in the grimy overgrown town. But the railways and cheap fares brought the possibility of at least day trips and excursions. Speeds of 60 m.p.h. and Gladstone's 'parliamentary trains', with fares not exceeding 1d. per mile, opened the coast resorts to the workers of Lancashire and Yorkshire. The Great Exhibition of 1851 was visited by over six million people, many of whom must have come from the North and the Midlands.

Within the towns, horse-drawn omnibuses and tramways multiplied until the coming of electrical power, when the great cities electrified their tramways, beginning with Leeds in 1891 and closely followed by Bristol, Glasgow, Hull and Liverpool. For medium and long distance travel the supremacy of the railways was undisputed; the first challenge did not appear until the end of the century. In 1887 came Daimler's first light-engined motor car; in 1895 the first motor exhibition was held in London.

The law restricting speeds of mechanically powered vehicles on the road to 4 m.p.h. in the country and 2 m.p.h. in towns and their attendance by a man with a red flag was repealed in 1896; in 1910 the London General Omnibus Company ran its first motor bus.

In the meanwhile the population of Britain had grown rapidly and continuously to 37 million in 1901. Sanitation and housing conditions had been improved by a succession of Public Health Acts culminating in the great Act of 1875. The worst horrors of the early Industrial Revolution had gone. Streets were paved, sewage and water were provided; gas, and later, electric lighting, was introduced. Every family was given the right to a frugal minimum of light and air. But the communal spirit of the town had been gravely damaged in the upheaval; the Georgian notion of urbanity had gone.

In this urban upheaval towns had become too large to comprehend as a unit; internally they were distorted and dismembered by a network of railway lines. The main roads were no more than widened lanes which formerly ran out into the countryside, and between them was a pattern of 'bye-law streets'—parallel lines of featureless cells providing the minimum amenities required by the law. The idea of the town as a good place to live had gone; that it might be *made* a good place to live was seldom contemplated. Instead, in revolt from the amorphous dreariness of the industrial town, the building of 'Garden Cities' captured the public imagination. We shall see later what Ebenezer Howard's proposals in 1898 really were, for they were only partly understood; but some of his garden city ideas were taken up as an easy way of escape from cheerless urban conditions. And so the new suburbs began to be built, agglomerations of little houses well set back from the road and with a garden at the back, mostly semi-detached, hardly ever in blocks of more than four. The country was to be brought into the town, and the more countrified it could be, the better. After the First World War a powerful stimulus was given to 'housing for the working classes'; building by councils and by speculators continued until the outbreak of the Second in 1939, even though it never measured up to the need.

By the 1930's the fresh pattern, if it can be called a pattern, had become established. The urban centre was still the place of administration of commerce and manufacture; around it was still the grey network of Victorian streets, to an increasing extent occupied by the lower-paid workers as their betters moved away to the outskirts and beyond.

UNIVERSAL SUBURBIA

The town had 'exploded'; suburbs had been added to the edges of the older built-up areas. Ribbon development was strung along the main roads; nearby small towns and villages had been swallowed up or invaded as outlying dormitory areas. Insistence on large gardens and low densities–at the very most twelve houses per acre–had increased the area of land in residential use almost beyond belief. 'Density zoning' at four, six, eight, ten and twelve houses per acre under the early planning schemes gave a legal sanction to the segregation of classes which had grown up. In the old towns there had been streets of large houses and streets of smaller ones. But they were closely associated; there was no wholesale segregation. In the new suburbs there were great Council estates housing immense working class populations; there were also great tracts of speculative housing for the lower middle classes. Communities grew up, as large as fair-sized towns, almost exclusively of people of one class. But they had no real centres which might have created or unified a civic life. Neither land nor funds were set aside for public buildings except schools (and sometimes not even these); little open space was allowed for and the provision of shops was haphazard. In these conditions community spirit languished.

Architecturally, the results were appalling. The Council houses were sometimes well designed internally but monotonous in the mass. In private building all emphasis was on individuality, on 'differentness', on privacy, however absurd this might be in most circumstances. About 1929 D. H. Lawrence wrote of English towns as 'a great scrabble of petty ugliness over the face of the land'. 'The English', he said, 'are town birds through and

1. The Grand Canal, Venice: St. Mark's Square and the main
waterfront; an example of the natural separation of transport
from pedestrians.

2, 3. The Venetian versions of (*top*) a pedestrian flyover and (*bottom*) a car park.

through. Yet they don't know how to build a city, how to think of one or how to live in one. They are all suburban, pseudo-cottagy and not one of them knows how to be truly urban.'[1] On the evidence at the time this could hardly be refuted.

It was in this period that the phenomenon of personal mobility first appeared—a phenomenon which is as fundamental in its implications as the coming of railways. It had been foreshadowed by the development of the bicycle in the 1880's but, though this inoffensive machine produced consternation at the time, it intro-duced no really serious problem. The cheap mass-produced car and motor cycle was a very different matter. For the first time anyone who could raise a few pounds was independent of public transport over considerable distances; he could take out his family into the country at weekends or the seaside for holidays. For the first time thousands, and soon millions, of these vehicles were crowding the roads designed for horses and carts. For the first time 'Everyman' required a garage—the equivalent of the coach-house and stable of the Victorian villa. And, for the first time, appreciable quantities of raw materials and manufactured goods were carried by road. Death and injury, traffic blocks, congestion and delay began to grow steadily and alarmingly.[2]

It is true that the railways maintained their sway well into the 1930's. To take a special case, London grew immensely because of the extension of the suburban lines into the countryside. Indeed the London Passenger Transport Board, which now con-trolled all bus services as well as the underground lines, did more to direct (or misdirect) the urban growth of London than any other agency was able to do. On the north side of the Thames, lines were constructed twenty miles out into virgin districts in the sure knowledge that houses would be built in and around exist-ing small towns and villages and provide passengers. On the

[1] D. H. Lawrence, *Nottingham and the Mining Countryside*.
[2] As early as 1903, an engineer, Lt. Col. R. E. B. Crompton, in a paper read before the British Association, advocated central traffic authorities, liaison between private and public transport, pedestrian precincts, under-ground garages and flyovers.

south side of the river the Southern Railway did much the same; the railways were supplemented by the Green Line buses of the L.P.T.B. The whole operation was backed by skilful and highly decorative advertising. The houses were built–both by speculators and the County Council–and the traffic did increase–from fifteen journeys per head per annum in 1860 to 456 on a much greater population in 1929, it has been estimated. By 1933 the Transport Board alone had 71,000 employees engaged in carrying the people of London from homes to work and in search of recreation.

Over the rest of the country cheap rail fares abounded, particularly for 'excursions', but a more significant factor was the development of motor bus services bringing to life again after their hundred years rest the intricate network of main roads and country roads and carrying men to their work, women to the shops and children to school.

Since the end of the Second World War the transport revolution has taken an even sharper turn. Every kind of road transport has multiplied at an astonishing rate; at a time when there is more movement than ever before, the railways carry a decreasing proportion of passengers and freight. As Professor Colin Buchanan has said: 'The revolution which has been overtaking us is that the motor vehicle has become an article of everyday use, a substitute for legs for walking and arms for carrying. People want motors everywhere and the real problem seems to to be arrange not for their diversion but for their peaceful penetration to every part of the town.'[1]

A rational pattern for the transport of Britain is long overdue. But even if we attained a unified national transport system in which railways, roads, canals and air all played their full part there would still be an enormous number of vehicles on the road. Motorways are essential, by-passes of towns are essential; but at the end of a journey vehicles need to get into towns. And within towns there is a complex internal traffic which is not in the least relieved by major roads and by-passes. Much of this internal movement is a function of our lack of planning–the haphazard

[1] C. D. Buchanan, *Mixed Blessing*.

siting of public buildings, industry, shops and offices, the design of residential areas as if the motor car did not exist. It could be greatly reduced by better arrangement. But even in a thoughtfully planned town we must expect a good deal of internal traffic. We have, in fact, to reconcile two conflicting needs—accommodation and circulation—the need to stay in one place for periods of work, play, rest and sleep, and the need to move about from home to work, school and shop.

In our more stationary moments, at home and at work, noise, fumes and vibration of traffic are nuisances which we could well do without; our need to move from place to place induces a further conflict between the desire of man-in-car to get somewhere else quickly and the slower movement of man-on-legs over shorter distances and his desire to stand and stare. This problem of living with the motor car is universal; it is most acute in Europe and America. The extreme case is probably to be found in Los Angeles, where there are six million people and three million cars. The city and its suburbs sprawl over an area 40 per cent greater than New York, which has four times the population; there are 650 miles of 'freeways', to which by 1980 are to be added 550 miles of 'expressways', apparently a super version of limited access highways.[1] This could be the shape of things to come elsewhere.

The only place in the world which has so far any full-scale answer is the ancient city of Venice, which excludes cars altogether, shuts them up in a vast garage at the end of the causeway from the mainland, and enjoys complete segregation of public and private transport on its network of waterways, whilst the pedestrian goes about his affairs in safety on land (Plates 1, 2 and 3). And even Venice has been threatened in recent years by schemes to take motor traffic across the islands of the lagoon and along the edge of the centre of the city.

Our personal lives are so influenced and invaded by the car that one is tempted to re-define the meaning of 'autocracy' as a society dominated, not by an autocrat, but by the automobile. With the world-wide growth of the car industry the problems we

[1] E. Higbee, *The Squeeze: Cities Without Space.*

31

face will shortly become equally serious in many lands which at present are in relatively early stages of development. But chaos and the car need not be synonymous; as the next two chapters will show, a number of ideas for eliminating conflict have emerged, even if we have so far done little to put them into practice.

CHAPTER 2

Prophets and Pioneers

The degradation of the town in nineteenth-century Britain was not universally accepted as the inevitable price of material prosperity–prosperity, that is, for the masters and the property owners. A steady and bitter fight went on for the improvement of housing and working conditions; here and there one finds visions of a more humane environment. It was indeed a period rich in Utopias–among them Robert Owen's projects for model industrial villages in Scotland in 1816, Titus Salt's modest Saltaire, built in the 1850's, and the technically more advanced 'Victoria' proposed by J. S. Buckingham in 1848, which was never carried out. These paved the way for Cadbury's Bournville (1879) and Lever's Port Sunlight (1888), where benevolent industrialists reproduced the rural atmosphere of cottage and village green in housing for factory workers. And in 1898, in his book, *Tomorrow*, Ebenezer Howard gathered together ideas from these and other sources and gave them a compellingly romantic form.

Garden City and Garden Suburb

Over the past sixty years Howard's proposals have been curiously distorted as much by his supporters as by his critics. Even now the words 'Garden City' suggest to most of us three things only–a self-contained town of limited size, large gardens and low density residential development. The first of these is correct so long as it is remembered that Howard proposed 'a cluster of cities . . . grouped round a Central City (so that) each inhabitant of the whole group, though in one sense living on a town of small size would be in reality living in, and would enjoy

33

all the advantages of, a great and most beautiful city' in which could be found the university, art galleries, theatres and so on which no small town can afford.

The second popular conception of 'Garden City', low density, was no part of Howard's original scheme. He specifically says that the average size of building plot would be 20 ft. by 130 ft. which works out at nearly seventeen houses per acre–well above the density of most housing of recent years including most of the housing in the new forms. Moreover, this simplified picture omits entirely some of the most original and striking items in Howard's programme, which relate to the detailed design of the residential districts. His model town (diagramatically shown by him with the proviso that it must be modified to suit the conditions of the site) is circular in plan and is bounded by a railway line linked to the main line. From the central park radiate boulevards 120 ft. wide dividing the residential zone into six sectors through which an 'annular ring' of parkway 420 ft. wide provides local open space and sites for schools and churches. There are several significant points here. Howard evidently felt the need to subdivide even a town of 30,000 population and arrived at a unit, or 'ward', of 5,000; he said, in fact, that each ward 'should be in a sense a complete town in itself' (Fig. 1).

Even though the motor car was not in common use, and is nowhere mentioned in Howard's book, the radial roads form the boundaries of the wards. It was intended that each ward should contain a cross section of all types of people; the selection of the word 'ward' implies that it should also be a local government unit. Howard stated emphatically that 'children must be educated near their homes' and placed a school centrally in each ward; in the early stages of the town's life he proposed that they should be used as churches, meeting halls and libraries.

One of the earliest and, in many ways, the best practical expression of Howard's ideas is to be found in Hampstead Garden Suburb. The original concept of its founder–Dame Henrietta Barnett–in the early years of this century was to demonstrate how thousands of people of all classes of society, of all sorts of opinions, and all standards of income, could live in helpful neighbourliness. The Garden Suburb was started in 1907 to the design of Sir Raymond Unwin; it was never able to cater for all classes; the site was far from ideal and too far from industries providing employment for 'artisans'. It has indeed always been a middle-class stronghold, but community life has been fostered by

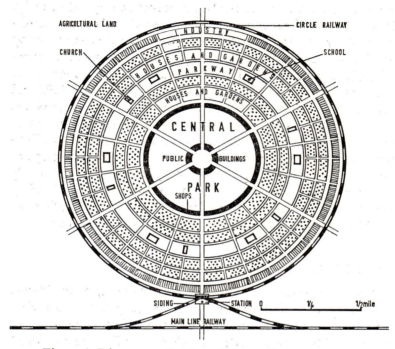

Figure 1. Diagrammatic plan of Howard's garden city.

the focus provided by churches and institute. It is a place of character and individuality, a kindly and intimate environment for children and family life, a striking contrast to the bye-law street pattern, and in a very real sense the pioneer example of the planned residential neighbourhood.

How fully Unwin absorbed and developed the ideas of Howard and the Barnetts can be studied in the pages of his great book *Town Planning in Practice*, published in 1909 and still a classic. He emphasized that, in the design for any but the smallest residential area, provision should be made for 'churches, chapels, public halls, libraries, institutes, shops, inns and schools' and that these should be 'the centre of the scheme'. This was impossible, he maintained, when designing for individuals or speculative builders; the 'creation of a village community' could only be achieved if control was in the hands of some co-operative body. He condemned the segregation of classes as 'foreign to the traditions of our country', resulting in bad municipal government by unfair distribution of local taxation, in exaggeration of differences of habit and thought, and architecturally, in dreary monotony.

But at Letchworth, the first garden city initiated in 1903 and designed by Unwin and his partner Barry Parker, some of the elements of Howard's proposals do not appear; there is no real attempt at subdivision into residential units. Here and at Welwyn, started in 1919, growth was slow; even by 1946 the population of the two towns together was only 40,000; this is probably the reason why the need for separate 'wards' and subsidiary social centres was not very obvious and one of Howard's most significant proposals was forgotten even by his followers.

THE LINEAR CITY

Also towards the end of the last century a completely different and highly revolutionary theory of town design had been devised by a Spanish engineer, Arturo Soria y Mata. In 1882 he advanced the idea of the 'linear city': 'a single street of 500 metres width and of the length that may be necessary–such will be the city of the future, whose extremities could be Cadiz and St. Petersburg, or Peking and Brussels. Put in the centre of this immense belt trains and trams, conduits for water, gas and electricity, reservoirs, gardens and at intervals buildings for different municipal services–fire, sanitation, health, police, etc.–and there would be resolved at once almost all the complex problems that are produced by the massive populations of an urban life. Our projected city unites the hygienic conditions of country life to the great capital cities and moreover assumes that the railways, like today's streets and pavements, will carry free or for little all citizens.'

Soria intended that his 'linear city' should connect old 'point cities' and envisaged continents crossed by great webs of these strip settlements. In fact he obtained a licence for a ring railway well outside Madrid and designed a linear community 55 kms. long. Only a quarter of this was ever built and the company which Soria founded ran into financial difficulties and was finally

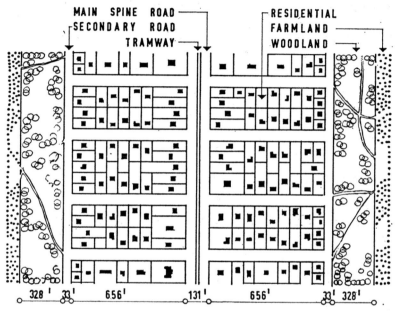

Figure 2. Soria y Mata's linear city pattern.

dismembered about 1940. But this was the first real response to the problem of integrating modern transport systems in the design of a town. It will be seen from the illustration that Cuidad Lineal consists of a central spine road and tram or rail track; on either side of this is a residential zone 200 metres wide served by transverse roads. Outside this is a subsidiary road and, beyond, woodlands and farm land (Fig. 2).

Cuidad Lineal is not a complete concept; at the ends of the line the problems of the 'point-city' remain unsolved. All the same it is a highly intelligent application of public transport to the service of the residential areas; it was not evolved for the age of the motor car but could have been adapted to it.

Linear City has never really been tried out in practice. It is true that 'ribbon' development has been an all-too-common feature, universally condemned as being extravagant in cost of services, in throwing purely domestic traffic on to roads intended to serve as by-passes or main routes, in devaluing agricultural land and in denying the traveller any view of the countryside through which he passes. Ribbon development—one plot deep along main roads—is, of course, not the same thing as Soria's linear town and loses all the merit that his idea had. But the revulsion from it has precluded any really serious consideration of Linear City. Not until the Second World War did Soria's theory reappear in an important form; it was one of the bases of the Mars Plan for London, as we shall see later.

THE NEIGHBOURHOOD IDEA

The work of Howard and Unwin, however, quickly took effect not only in Britain but, in a different way, in the United States, where in the early years of this century the 'Community Centre Movement' grew up. Based on the principles of Canon Barnett's Toynbee Hall, this movement aimed to 'animate civic life by providing a common local meeting place'. It was the origin of the proposal to use schools for communal activities, after school hours and during summer months, reported in 'The Wider Use of the School Plant' by Clarence A. Perry in 1910.

The words 'neighbourhood unit' seem to have been used for the first time in connection with a planning competition in Chicago about 1916; the first full statement of the idea appears in Clarence Perry's monograph in Vol. 7 of the *Regional Survey of New York*[1] published in 1929. From 1912 Perry had lived at Forest Hills, a suburb of New York which had grown to a population of 5,000 and had a rich variety of local societies and social gatherings which seemed to him the hallmark of a healthy

[1] *Regional Survey of New York and its Environs.* Vol. 7. 'The Neighborhood Unit: a scheme of arrangement for the Family Life Community.'

community. It fell short of his ideal, however, for several reasons: it lacked physical definition, it had no community centre and not enough shops or open space, and its residential streets had also to carry through traffic. The 'neighbourhood unit' theory which Perry put forward aimed at eliminating these failings. 'Our investigations,' he said, 'showed that residential communities, when they meet the universal needs of family life, have similar parts performing similar functions. In the neighborhood unit system those parts have been brought together as an organic whole. The underlying principle of the scheme is that an urban neighborhood should be regarded both as a unit of a larger whole and as a distinct entity in itself. There are certain other facilities, functions or aspects which are strictly local and peculiar to a well-arranged residential community. They may be classified under four heads: (1) the elementary school, (2) small parks and playgrounds, (3) local shops and (4) residential environment. Other neighborhood institutions and services are sometimes found, but these are practically universal.' He laid down the fundamental elements on which he intended the neighbourhood unit should be based – size, boundaries, open spaces, institutional sites, local shops and internal road system. It is worth while examining each of these in some detail (Fig. 3).

Perry's unit was based on the population ordinarily required for one primary school. The area of the unit would depend on population density but the factor of distance was considered. The school was to be placed at the centre; the desirable maximum distance for a child to walk was put at half to three-quarters to a mile; this set limits on the size of the unit. Perry also discussed desirable size from the point of view of residential characteristics; this he found more difficult to decide but thought that 'character expressed in local associations' seemed to be strongest when representing about 1,000 families or a population of 5,000. A theoretical study of Queen's Borough, N.Y. by Robert Whitten attached to Perry's monograph provides for a population of 6,124 including 1,021 children of school age. The total area is 160 acres of which 90 acres is purely residential.

The neighbourhood was to be bounded on all sides by arterial roads, wide enough to serve all through traffic. This emphasis was no doubt brought about by the grid-iron street pattern almost universal in the U.S.A. Perry was much concerned that the boundaries of neighbourhoods should be clearly defined and

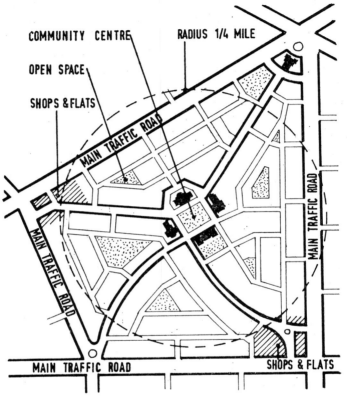

Figure 3. The neighbourhood unit as proposed by Clarence Perry in 1929.

saw busy traffic avenues as forming obvious and effective limits for residential areas.

A system of small parks and recreation grounds was to be provided amounting to about 10 per cent of the total area of the unit. The greater part of this was to be at the centre and attached to the school, which was also to be used as a community centre;

small children's playgrounds were to be dispersed throughout the unit so that no child need walk more than a quarter of a mile to one of them.

Shops were to be provided on the edge of the unit, preferably at traffic junctions and adjacent to the shops of adjoining neighbourhoods. Perry's argument in favour of this arrangement was that a cluster of shops serving, say, four neighbourhood units offered wider choice and competition. He felt that shops at the centre of the unit would draw commercial traffic through residential areas and that their presence damaged residential quality.[1]

Perry's fourth major objective was a special street system; main arterial streets were to be designed to carry heavy through traffic and this was to be discouraged within the neighbourhood unit. Here the purpose of streets was to make local circulation easy; even after eliminating through traffic, pedestrians should be segregated from moving vehicles, and underpasses constructed where necessary.

It is abundantly clear from both the text and illustrations of Perry's works that he had been deeply influenced by the sociological ideas of Howard and Unwin. He went even further in urging that this theory was applicable not only to new districts but to areas already built up. Here he thought the neighbourhood pattern 'could bring the local community into relief and enable its residents to see it as something apart from the rest of the city'.

THE RADBURN IDEA

A development of the neighbourhood unit pattern, introducing new features aimed at solving the traffic problem, appeared in the planning of Radburn, New Jersey. Here the City Housing Corporation, a private company who had already built Sunnyside Gardens, New York between 1924 and 1928, acquired about two square miles of irregular land. It was their intention to build

---

[1] Clarence A. Perry, *Housing for the Machine Age.*

an American 'Garden City' with a population of 25,000 in three neighbourhoods.[1]

The plan of the town, prepared by Clarence Stein and Henry Wright, is illustrated in Fig. 4: 'Neighborhoods were laid out with a radius of half a mile centering on elementary schools and

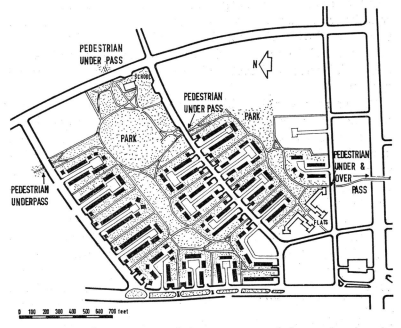

Figure 4. Radburn: a part of the town as carried out, showing the open space and footpath system.

playgrounds. Each was to have its own shopping centre.'[2] The size of the neighbourhood (7,500–10,000) was determined by the number of children to be cared for by a single school. The main educational and cultural centre for the town was placed equidistant from the three proposed schools, within a mile of all the houses, and to it was attached a central recreational area. The commercial centre was located west of the cultural centre, close

[1] C. Stein, *Toward New Towns for America.*
[2] Ibid.

to the proposed state arterial road at the main entrance to the town.

The general planning of the neighbourhood units closely followed Perry's ideas. It was based on the school as a community centre; through traffic was canalized on the main roads of the town, and shopping centres were placed on these roads. The detailed planning of the residential quarters was, however, the most striking feature.

The traditional grid-iron pattern of streets, all equally attractive to through traffic, was abandoned and a number of new elements introduced. A logical system of specialized one-purpose roads was devised—arterial roads linking with the surrounding area, main town roads, main estate roads and service roads for houses. The main estate roads enclosed 'superblocks'—areas of 30 to 50 acres within which were the houses and the culs-de-sac which served them. Large areas of open space were left in the centre of the 'superblocks' providing a backbone of continuous park, towards which the houses faced, and through which ran footpaths; underpasses and footbridges across traffic roads linked the 'superblocks' together. There was in fact to be complete segregation between the pedestrian and the motor car.

Radburn was never completed. Almost as soon as the first inhabitants had moved in, the slump hit America. Building and the purchase of land went on at a diminishing rate until 1933 when the Corporation was forced into bankruptcy and had to sell the greater part of the undeveloped land. 'The dream of a complete new town had been destroyed by the Depression.'[1]

On the surface Radburn may appear to have been a failure. There was not an adequate surrounding area to provide a protecting green belt so that other and less satisfactory building grew up on its boundaries. Industry did not develop, partly because of the economic depression, partly because of poor communications. And from the first the Corporation had to contend with the fact that Radburn was never a separate local government entity.

But the essential principles of the Radburn idea were achieved in the small segment that was built and are a great success: 'Those

[1] C. Stein, *Toward New Towns for America*.

who live in Radburn and have lived there for any length of time find it has served its objective of making home and community life more reposeful, pleasant and safe–and particularly safe for children. The physical plan of central parks, superblocks without through traffic, safe walks, houses facing on gardens and parks along with the convenience of service here have, they find, given them a quality of living that, as medium income folks, they could not find elsewhere. . . . One is conscious of a sense of the stability

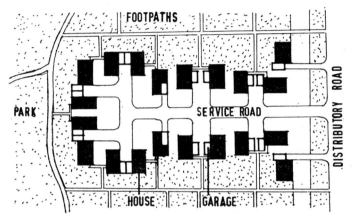

Figure 5. Radburn: a detail of one of the lanes.

of an old community with a definite character and roots . . . an awareness of neighbours and their interests, of their abilities, their ambitions and their needs.'[1] Even though Radburn had to accept the subsidiary role of a small suburb it splendidly exhibits the validity of the neighbourhood idea in the United States.

At Baldwin Hills Village, Los Angeles, in 1941, another and more comprehensive project based on the 'Radburn Idea' was begun. The site available was only 80 acres on which a pattern of continuous through streets had already been planned by municipal engineers. After a lengthy struggle all through-streets except one were eliminated; the one which remains does, how-

[1] C. Stein, *Toward New Towns for America*.

ever, segregate the Village from its shopping centre at the cost of serious traffic risks.

With this exception the Radburn 'superblock' pattern has been followed: culs-de-sac lead from the peripheral main roads to garage courts at the backs of houses and flats, whilst in the centre of the superblock is generous and well planted open space. The garage court is a development from Radburn; within it is a garage for each house, parking and manoeuvring space, public group laundries and outdoor but enclosed drying yards; each

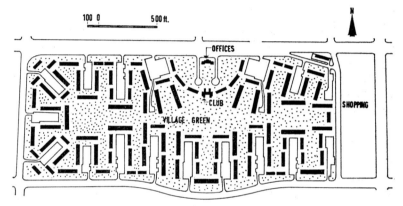

Figure 6. Baldwin Hills Village: the scheme as completed.

court serves less than fifty families. The garden courts at the front of the houses, 100 or more feet wide, are entirely open and unfenced although a strip of about 20 feet is regarded as private to individual houses. These garden courts connect with the central park area which, although only 250 feet wide, has a surprising sense of space due partly to the limited two-storey height of surrounding buildings. This area is open for recreational use of all kinds; in practice small children mostly play in the garden courts near their homes (Fig. 6).

Strangely enough, on either side of Baldwin Hills Village, the typical speculator's estate layout has gone on undeterred, with meaningless winding corridor streets, no open space and a wastefully low density of building.

45

NEIGHBOURHOODS IN THE U.S.S.R.

The growing acceptance of the neighbourhood idea was by no means confined to English speaking countries. In Russia, for example, in the 1930's, there seems to have been competition in proposals for Moscow between the 'linear city' schemes of Gins-berg and the 'neighbourhood' schemes put forward by Ernst May, a former pupil of Unwin. In 1936 Dr. Eugene Kaufmann, formerly Director of Housing at Frankfurt-am-Main, gave the entire credit for the neighbourhood idea to Russian planners.[1] He referred to the translation of a paper produced in Moscow in 1931 by Goldenburg and Dalganoff, which argues the case for communal living on political and economic grounds. The result is a pattern curiously similar to Perry's, with main roads for heavy transport only and shops, schools, creches, restaurants and so on planned for the needs of a hierarchy of units. The smallest 'collective' was to be a group of five-storey flats housing 1,000–1,200 people, junior schools would serve two 'collectives' and senior schools would accommodate 1,000 pupils. Open space was to be provided on the basis of 10 sq. metres per person, not including commons and parks. Generally the aim would be, by caring for children in creches and schools and by eliminating private cooking and laundry, to free women for 'socially useful work for the whole community'. Kaufmann illustrated his paper with diagrams from Russian sources including a drawing of the smallest unit (eight blocks of six flats served by a communal wash house, a kindergarten and a creche), another of a neighbour-hood of 10,000 made up from such units, and a plan of the city of Magnitogorsk laid out on these lines.

It seems possible that Kaufmann himself was not aware of the work of Perry, Unwin and other British and American planners, although he did refer to a book by A. P. Simon,[2] a Manchester City Councillor, actively concerned with housing, in which neighbourhood principles were advocated.

[1] 'Neighbourhood Units as New Elements in Town Planning', *R.I.B.A. Journal*, 19th December, 1936.
[2] A. P. Simon, *Manchester Made Over*.

PROPHETS AND PIONEERS

Planning Practice before the Second World War

Even as late as 1939 the only theory of planning generally accepted in Britain was Garden City, a formula for escape from the 'dark satanic mills' and a fresh start in green fields. The details of Howard's project show a sensitive approach to the fulfillment of men's needs as individuals and as social and political beings. But this theory was not put into general practice; Letchworth and Welwyn grew only slowly and no other new towns were attempted. Nor was any effort made to develop Howard's theory, evolved before the general use of motor cars, to cater for the growing traffic problem.

Instead, Garden City degenerated into the sentimental and unrealistic form of garden suburb, a bastard version of some of Howard's principles. Between 1919 and 1939, a million houses were built by local housing authorities and three million by other agencies which, in the main, means speculative builders. They were almost all semi-detached or detached houses at low densities with roads laid out to geometrical patterns which were only comprehensible from the air or which curved unmeaningly 'in the interests of amenity'; building lines were invariably parallel with the streets and often very deep. It as a period of intense individualism. Seldom were new public buildings, or sites for them, provided and even shops were located in the most haphazard fashion. No attempt was made to face the growing traffic menace or to foster or recreate community values.

The neighbourhood unit theory had attracted little attention. In an article on educational buildings[1] Wesley Dougill referred to Perry's work and urged that greater thought be given to the neighbourhood unit principle. Whether we liked it or not, he argued, traffic was dividing the town into a number of cells; the school, from being a cast-off, was raised by Perry's theory to be a most important element; community buildings at the centre of the unit would help to galvanize it into an entity. Dougill pointed out that the City of Nottingham had been divided into sixteen

[1] Wesley Dougill, 'Educational Buildings', *Town Planning Review*, July, 1934.

districts for elementary school purposes; at the Apsley House Estate a group of schools was to be erected at the centre which would serve as community buildings in the evenings, at weekends and in the vacations.

In a later article[1] Dougill pointed to Wythenshawe, on the outskirts of Manchester as a more complete example of the neighbourhood unit in England. The Wythenshawe Estate was acquired by the Manchester Corporation, its 5,567 acres were incorporated in the City in 1930 with the object of accommodating people employed in Manchester in a satellite town. Here, the neighbourhood pattern was used from the beginning by the architect and planning consultant, Barry Parker, a brother-in-law and partner of Sir Raymond Unwin. The site was divided into large sections bounded by traffic routes, each section having a school near its centre. In the location of local shopping groups American practice was adopted; they were placed at main road crossings at about three-quarter mile centres. By 1938 over 30,000 people had been housed at Wythenshawe and light industries were growing up on the trading estate. With these and a few other exceptions the neighbourhood idea went unnoticed.

The growth of the acceptance of the neighbourhood unit idea may be studied in James Dahir's bibliography.[2] Yet even in the United States it was still a theory, only partly put into practice and developed at Radburn, Baldwin Hills, and in the Green Belt Towns, which were little more than single neighbourhoods in themselves; it was far from being accepted as general practice. In Russia the theory seems to have been used on a considerable scale but contacts between planners of East and West have been so poor and infrequent that too little is known about Soviet experience.

It is evident that a pattern for living in the twentieth century had not yet crystallized. Neither Garden City nor Neighbourhood Unit principles went any way to solve the growing problems of congestion and decay in central commercial and industrial districts or of the overall circulation of increasing volumes of traffic.

[1] Wesley Dougill, 'Wythenshawe', *Town Planning Review*, June, 1935.
[2] James Dahir, *The Neighborhood Unit Plan: its Spread and Acceptance*.

These were still regarded as natural features like the weather which one could grumble about but must accept as inevitable.

LA VILLE RADIEUSE

A complete theory of city planning had, however, been worked out some years before by the Swiss architect, Le Corbusier. In 1922 his scheme for a contemporary city was shown for the first time; it was 'greeted with a sort of stupor; the shock of surprise caused rage in some quarters and enthusiasm in others'.[1]

Both the plan itself and the book *Urbanisme*, which Le Corbusier later wrote to explain and amplify it, are outstanding landmarks in the history of modern planning. More vividly and completely than anyone else he pointed out the defects of twentieth-century towns. 'A town is a tool' he wrote 'towns no longer fulfil this function. They are ineffectual, they use up our bodies, they thwart our souls. The lack of order to be found everywhere in them offends us; their degradation wounds our self-esteem and humiliates our sense of dignity. They are not worthy of the age; they are no longer worthy of us.' The root of the problem, he argued, lay in the increase in traffic – 'its power is like a torrent swollen by storms . . . all our being is absorbed in living like hunted animals . . . the norm of our existence is completely demolished and reversed'.

The basis of Le Corbusier's thought was not to try to overcome existing conditions but to construct 'a theoretically watertight formula to arrive at the fundamental principles of town planning'. The requirements he laid down were that the centres of cities must be de-congested to provide for the demands of traffic, density in the centre must be increased to allow for the close contact demanded by business; there must be improved facilities for getting about; and the area of green and open spaces must be increased. The satisfaction of these four apparently irreconcilable demands was bound to assume revolutionary forms.

Le Corbusier assumed a population of three million inhabitants. These he classified as citizens, who lived and worked

[1] Le Corbusier, *The City of Tomorrow* (translated by F. Etchells).

49

in the city, suburban dwellers, who worked in the outer industrial zones and lived in garden cities, and a third type, who worked in the city but lived outside. The city itself, as the business centre, should, he thought, be compact, lively and concentrated so as to reduce the distances to be covered; fresh air and quiet were essential; therefore, open space must be ample; the only solution was to build vertically which meant the end of the traditional 'corridor street'.

The basis on which Corbusier founded his proposals was a national transport and traffic system. He proposed three different classes of roads, one, at the lowest level, for heavy traffic collecting and delivering goods, above these the network of ordinary access streets, above again, on concrete viaducts, the two great axes of the city—arterial roads for one-way traffic linked at half mile intervals to the lower systems. The subsidiary street pattern would be a grid-iron at 400-yard intervals, thus reducing the number of cross roads to a minimum but not exceeding the distance which could conveniently be walked.

A very complete railway system was also proposed. There was to be an underground station in the centre of each block of 400 yards square with a tube network serving the city and the main arteries, below it one-way loop systems for suburban traffic, and below again four main lines running north, south, east and west. All these systems would connect in one great central station on the roof of which would be an aerodrome.

Around the station, in the heart of the city, Corbusier proposed a group of sixty-storey office buildings set in a great park in which would also be restaurants, theatres, shops; adjoining this would be the public buildings, museums and administrative offices; beyond again would be parkland which might later be required in part for extension of the city centre. Around the central area were to be the residential quarters laid out within the 400-yard grid-iron road pattern and all in the form of flats. The industrial quarters and goods stations were to be separated from the city by open land which extended all round it; outside this green belt would be the garden cities which would house two-thirds of the residents of the city complex of three million people (Fig. 7).

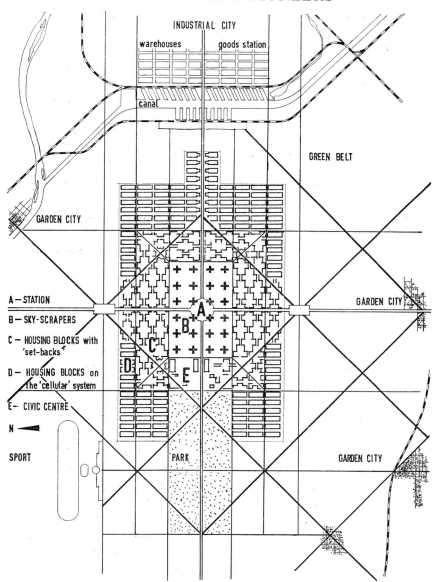

INDUSTRIAL CITY

warehouses — goods station

canal

GREEN BELT

GARDEN CITY

A — STATION

B — SKY-SCRAPERS

C — HOUSING BLOCKS with 'set-backs'

D — HOUSING BLOCKS on the 'cellular' system

E — CIVIC CENTRE

N

SPORT

PARK

GARDEN CITY

GARDEN CITY

Figure 7. La Ville Radieuse: diagrammatic plan of Le Corbusier's ideal city.

The whole pattern of Corbusier's city was logically based on modern mobility and his passion for geometry; a first impression might be that it was too geometric and mechanical. It was conceived as a 'laboratory experiment' on an ideal site which would have to be adapted to actual conditions. It was very far from being inhuman; its purpose was to free citizens from the hazards of traffic–'dynamite flung at hazard into the streets'; it provided a variety of different forms of living conditions; there was a high sense of the need for open space and fine environment in all parts of the city.

Corbusier's schemes for taking advantage of modern technical resources to build vertical 'garden cities' in the form of large long slab blocks of flats have had a deep influence throughout the world. The main prototype in his own works is the Unité d'Habitation at Marseilles started in 1945. This is an eighteen-storey block of 337 flats–'a box of homes' as Le Corbusier himself describes it. It is, in fact, more than that, for within it are a creche and a kindergarten, on the roof are a swimming bath and playgrounds for children, and for adults a gymnasium, running track and solarium. Within the $8\frac{1}{2}$ acre site are a garage, swimming bath and sports ground; a school and other requirements could not be provided within the funds available. This is not the place to consider whether the Unité is completely successful: in any case it is a single unit only of the 'Ville Radieuse' which Corbusier advocated for forty years, and should be judged as such.

The principles of 'La Ville Radieuse' have been generally dismissed as wild and impracticable. Certainly they were completely different from what had gone before; the difference between them and Howard's Garden City proposals, only twenty-four years earlier, is striking. They were in fact directed towards a completely changed world; as Corbusier saw, the onset of the motor age must radically alter our ways of urban life. We are only now beginning to discover that the pattern he suggested contains many of the clues to living with the motor car.

CHAPTER 3

Planning in the Forties

It is difficult to recapture now the atmosphere of enthusiasm for planning which was universal during the War when our cities were daily and nightly being attacked by high explosive and incendiary bombs. The ordeal was lightened by an intense belief that the towns so cruelly devastated would be rebuilt quickly after the War, that worn-out districts would be redeveloped and that many of the causes of ill-health, inefficiency and drabness would be rooted out. The widespread destruction was seen as an opportunity to rebuild – and build better.

In this atmosphere, theories about planning flourished and many schemes were put forward by individuals, groups, and official and semi-official bodies. It is worth while looking again at some of them not only because they have contributed to present-day practice but because, re-examined in the light of experience, they have still more to give.

THE PRECINCT

One of the most significant was the theory of 'precinct' planning put forward in 1942 by Sir Alker Tripp, then Assistant Commissioner of the Metropolitan Police.[1] He pointed out, as Le Corbusier had done, that road traffic conditions had rendered the all-purpose highway obsolete – a road could no longer fulfil the function of a main route together with that of an access to buildings along its frontage. He proposed clear definition of roads as arterials which formed a national network, sub-arterials which connected this network with towns, and local roads which should have limited access to sub-arterials and none to arterial

[1] H. Alker Tripp, *Town Planning and Road Traffic*.

roads. Under this system areas would be created, each of which would be served by a local system of minor roads devoted to industrial, business, shopping or residential purposes. Each area would be 'a centre of life and activity' to which Tripp gave the name of 'precinct'.

Tripp went on to suggest that no part of each precinct should

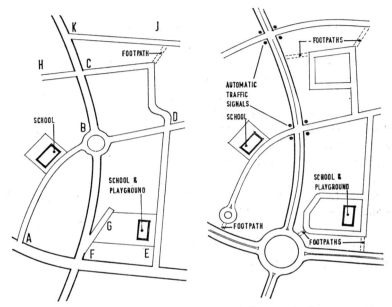

Figure 8. Alker Tripp's 'precinct': on the left, a typical street pattern with many dangerous traffic intersections: on the right, the 'precinct' principle has been applied to the same district.

be more than a quarter of a mile from an arterial or sub-arterial road which could carry a bus service. A practical application of his principles is illustrated (Fig. 8). The first drawing shows how a tract of land would normally be laid out by developers; the second shows the results of precinct planning, reducing the access on to main roads and designing these roads themselves for their function in carrying through traffic. This was, of course, no more than an English version of the Radburn superblock, evolved by a traffic expert looking at the problems produced by the motor vehicle from his own specialized point of view. But it was a most important contribution; it is remarkable that the differing functions of roads are only now beginning to be studied seriously.

THE VILL

At much the same time support for the neighbourhood theory came from another source which attracted little attention. In 1940 C. B. Fawcett, Professor of Geography in the University of London, put forward his own ideas for residential units. His studies of the distribution of settlements led him to the conclusion that technical and economic organization had got more and more out of step with social organization. 'The chief social weakness of town life', he wrote, 'is that . . . the individual tends to be lost in an anonymous crowd. . . . To have too many neighbours is to have none.' Residential units–to which Fawcett gave the name of 'vill'–should be based not on economic, administrative or architectural considerations but on social factors, the most important of which he thought was the primary school. Each vill would have a one-stream school (120 pupils), a community centre, a café and a public house; one secondary school would serve three vills. This system, Fawcett suggested, could be applied to new and redeveloped residential areas in towns and in the countryside. The concept was strikingly similar to that of Perry but did not appear to stem from it.

THE SOCIAL STRUCTURE OF TOWNS

A comprehensive survey of the needs of modern residential areas in Britain was published in 1943 by the National Council of Social Service.[1] Defects in the housing estates built after the First World War were pointed out: excessive travel between homes and workplaces had resulted in the expenditure of money, time and effort which militated against the capacity of workers to

[1] *The Size and Social Structure of a Town.*

develop the life of their new surroundings and retarded the growth of neighbourly relations; class differences had been emphasized by the segregation of income groups in separate residential districts, and this resulted in lack of leadership on new municipal estates; shops, libraries, churches, etc., were all too few; cinemas and public houses were the only 'community' buildings which could generally be found; very often no sites had been left for public buildings or shops to be built later; indeed it was sometimes difficult or expensive to provide sites for schools.

The report argued that future planning schemes, whether for redevelopment of bombed areas, enlargement of existing towns or creation of new urban units, should be based on 'neighbourhood units' each fitting into the town to which it belonged and each containing a socially balanced population. No standard size of unit was proposed; the report pointed out that size must depend on walking time, distance and population density taken together; the limits were thought to be 7,000–10,000 people and ten to fifteen minutes walking time from periphery to centre, where communal buildings would be needed.

Admirable though this report was it failed notably in its lack of reference to the problem of the motor vehicle. There was no real recognition of the effect of increase in goods vehicles or private cars and their penetration into all parts of the town; the emphasis was entirely on social structure.

OFFICIAL OPINIONS

The Dudley Report,[1] published in 1944, contained the report of a Study Group of the Ministry of Town and Country Planning on site planning and layout in relation to housing. It followed the lines of the N.C.S.S. report closely, recommending: '. . . the creation of residential neighbourhoods' which would be 'comprehensible entities' in themselves and might even be local government entities (wards). A maximum of 10,000 persons was suggested, with all houses within ten minutes walk of the

[1] *Design of Dwellings*: Report of the Sub-committee of the Central Housing Advisory Committee appointed by the Ministry of Health.

neighbourhood centre. The size of the neighbourhood in terms of acreage would vary, according to density, between 168 and 482 acres for 10,000 persons, with a full provision of schools, open space, shops, public buildings and service industry. The careful disposition of these amenities would go a long way to solving the problem of 'social balance' (families belonging to different income groups) but the main key to a full solution would lie in the grouping of various types of dwellings. It was suggested that minor groups of homogeneous character providing for 100–300 families offered the best hopes of success.

The open space system, said this report, should be closely related to the dwellings. A continuous pattern providing a system of safe pedestrian ways was recommended, although it was pointed out that open space on the perimeter would act as a useful buffer between the housing groups and traffic arteries or railways.

Even more definite official approval was given to the neighbourhood unit, as a basis for community structure and town planning, in the Housing Manual 1944, issued by the Ministry of Health. This, in effect, accepted the findings of the Dudley Report but suggested that neighbourhoods might be as small as 5,000; it also gave the warning that schools, shops and other communal buildings should be completed by the time houses were ready for occupation.

Side by side with plans for building and rebuilding residential districts the creation of complete new towns was being actively considered. In 1945 a strong Committee was set up under the chairmanship of Lord Reith 'to consider the general questions of the establishment, development, organization and administration that will arise in the promotion of New Towns in furtherance of a policy of planned decentralization from congested urban areas; and in accordance therewith to suggest guiding principles on which such towns should be established and developed as self-contained and balanced communities for work and living'.[1] The activities of the Committee were, therefore, not restricted to relieving the problems of London and the South-east but ranged over the whole question of New Towns anywhere in the United

[1] *First Interim Report* of New Towns Committee.

Kingdom. They emphasized, in their Final Report in July, 1946, that: 'Our recommendations do not imply that there should be any one standardized pattern of physical or social structure. What may suit the Londoner may not suit the Lancastrian or Clydesider. There should be full latitude for variety and experiment.'[1]

Two types of new towns were considered; towns that were entirely new, placed where previously there had been only a scattered and rural population; or major extensions of existing small towns. Whilst recognizing that there were cases when the second possibility would be practicable and beneficial, the Committee came down firmly in favour of new towns on relatively undeveloped sites. The optimum figure decided on was 30,000 to 50,000 in normal conditions. In towns of this size all dwellings could be within walking or cycling distance of the centre and workplaces, the countryside would easily be reached, and a sense of unity and local pride could be attained. The Committee pointed out that their terms of reference required the new towns to be 'balanced'. This was not only an economic issue but one based on social considerations. They, therefore, felt it important that: '. . . proprietors, directors, executives, and other leading workers in the local industries and businesses should live in the town and take part in its life. Many professional men and women, writers, artists and other specialists not tied to a particular location should find a new town a good place in which to live and work.' This diverse and balanced social composition was to be established at the beginning; if the first houses built were all of the minimum standard, the town would be stamped at the outset as a 'one-class' town and it would be difficult to redress the balance later, so that in the earliest development there should be some groups of houses of varying size as well as sites for houses to be built to the requirements of owner-occupiers. The Committee saw clearly that a sustained policy was needed to combat the tendency to segregation caused, not so much by differences in income, as of social background resulting in different social behaviour.

[1] *Final Report* of New Towns Committee.

PLANNING IN THE FORTIES

The recommendations of the Reith Report quickly took on the shape of reality for in his 'Greater London Plan', published in 1945, Sir Patrick Abercrombie proposed that half-a-million people from the Metropolitan area should be rehoused in ten new towns in Berkshire, Essex, Hertfordshire, Kent and Surrey. These were not to be dormitory areas but fully equipped communities including industry to provide local employment for a large proportion of the town's workpeople.

THE COUNTY OF LONDON PLAN

The principal landmark in the study of the design of residential areas during the War is undoubtedly the County of London Plan published in 1943. The authors, Sir Patrick Abercrombie and J. H. Forshaw, who were assisted by a distinguished team, mention in their report the important personal contribution of Wesley Dougill, whose appreciation of the neighbourhood concept has been referred to in an earlier chapter, and who died a few months before the Plan was published.

The report opens with a close analysis of the social structure of the capital which recognized that many of the residential communities can be traced back to the original villages which have been submerged, whilst others have been determined by geographical conditions or by man-made barriers such as railways, canals and concentrations of industry. It is pointed out how, for example, Bermondsey and Battersea suffered by the building of railways which cut across their centres. In other cases, such as New Cross and New Cross Gate, shopping centres and community buildings had grown up around the historic nuclei on the main roads and produced acute traffic congestion. 'This social structure is of the utmost importance in the life of the capital. Community grouping helps in no small measure towards the inculcation of local pride, it facilitates control and organization and is the means of resolving what would otherwise be interminable aggregations of housing. London is too big to be regarded as a single unit. If approached in this way its problems appear overwhelming and almost insoluble.' It was, therefore, proposed to 'emphasize the identity of the existing communities, increase their degree of segregation and, where necessary, to re-organize them as separate and definite entities'; at the same time it was stressed that segregation was not to be taken so far as to compromise interdependence.

To strengthen community structure, it was proposed that each community should possess its own schools, public buildings, shops, open spaces, etc. Care was to be taken to avoid new main traffic routes cutting across communities. Within the communities so conserved or created there were to be smaller neighbourhoods of between 6,000 and 10,000 related to the elementary school and the desirable walking distance from home to school: 'The communities would possess as great a variety in size as the existing ones . . . an objective would be to have them not so big as to make the individual feel overwhelmed nor so small as to make the units unable to support a reasonable equipment of schools, community buildings, shops, etc.'.

The Plan includes a large number of drawings (two of which are illustrated, Figs. 9 and 10) showing the application of the principles to theoretical and actual sites. The community centre, school and shopping area are centrally placed in each scheme; there is a mixture of houses and flats in varying proportions to give residential densities of between 100 and 200 persons per acre, i.e., not including areas allotted for open space and communal buildings; the corresponding gross neighbourhood densities were sixty and ninety-seven persons to the acre respectively. The neighbourhood plans are so designed as to eliminate through traffic from residential areas on the 'precinct' pattern.

The proposals for the improvement and rationalization of main road and rail communications were less radical than in the Mars Plan but they were bold and courageous; few of them have yet been carried out. The report as a whole, however, gave a tremendous impetus to positive planning throughout the country. Over the next few years plans appeared for other cities, following the lead which it had given, and adopting the same principles. During the same period plans were prepared by Thomas Sharp

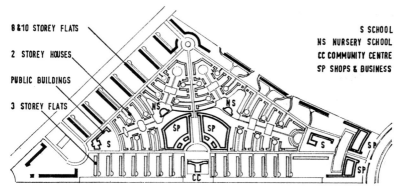

Figure 9. County of London Plan: theoretical study of a
neighbourhood at a density of 136 persons per acre.

for Oxford, Durham, Exeter and other towns. These were detailed
and humane studies, notable for their emphasis on preservation
of the character of these historic towns but at the same time for
clear-sighted attention to traffic circulation. Sharp's attitude was
that the wholesale widening of main streets was functionally
ineffectual as well as destructive of character; he proposed relief
roads which were rationally and ingeniously located. His contribution to thought on central area problems has never been

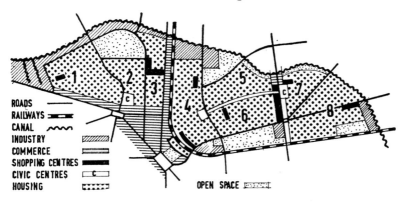

Figure 10. County of London Plan: redevelopment of an area in East
London: eight neighbourhoods with a population between 6,900 and
10,800: open space standard 4 acres per 1,000.

sufficiently recognized; his aesthetic opinions were acceptable to
the preservationists; the logic of his road proposals, which were
the vital corollary, was often unpalatable. The controversy in
Oxford about a road across Christ Church Meadow has
smouldered since 1948; various other solutions have been proposed and investigated but none of them put into effect.

THE 'MARS' PLAN FOR LONDON

Quite the most revolutionary of all the proposals made during
and immediately after the War were, however, those of the Town
Planning Committee of the Mars (Modern Architectural Research) Group which was made public in 1941. It made detailed
proposals for the social and cultural needs of the people of the
Metropolis, but is particularly notable because it boldly faced
the traffic problem and based its proposals on a logical transport
system (Fig. 11).[1]

A main railway arterial was proposed, roughly parallel with
the River Thames, and a ring of long-distance railway lines providing links with the provinces and circulation for goods traffic.
The suburban railway system was to be completely segregated
from this except at a limited number of interchange stations. The
pattern proposed for the road system was a spine road through
the centre, from which ran, north and south, parallel arteries
intended for public transport and cyclists only. The residential
areas were to be about half a mile wide either side of these
arteries and between them were to be broad green spaces. Private
cars were to use intermediate roads through the green spaces; the
number of people going by car to the centre would be relatively
small because the speedy and fully organized public transport
would provide for most needs. It was seen that the criterion for
traffic is time and not distance, so that if the proportion of turning
traffic could be reduced, public road transport could do more
journeys, and congestion would be relieved.

The residential pattern was based on units of 1,050 people, six

[1] A. Korn and K. J. Samuely, 'A Master Plan for London', *Architectural
Review*, January, 1942.

of which were combined to form a neighbourhood unit. Borough units of four to eight neighbourhoods were proposed, building up into districts of twelve borough units.

The Group did not claim that the plan was original in all its features; obviously much had been learned from Soria y Mata,

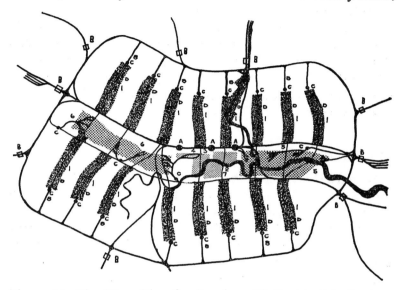

Figure 11. The Mars Plan for London. (1) Residential districts; (2) Commercial administration (City); (3) Political administration; (4) Shopping centre (the goods rails shown are underground); (5) Cultural centre and park; (6) Western industries; (7) Eastern industries and Port of London; (8) Local industries, possibly combined with satellite towns. A. Main passenger stations; B. Main goods stations; C. Secondary goods stations; D. Market halls. The map shows railway connections, but all roads are omitted.

Le Corbusier and others. They did urge however that London could be rebuilt on rational lines and the solution based on scientific investigation. The Plan was most original in its emphasis on communications and its insistence that increased use of the private car in the future could only be checked by rationalization of public transport systems. The Group pointed out that although the realization of the plan involved immense costs these would be much more than balanced by the saving in time, decreased freight charges, fuel costs and land values. The total cost, they estimated, equalled the amount spent by Great Britain in three months of the War; rebuilding on these lines could be achieved in twenty years. But the Mars Plan was quietly passed over as being too radical and too destructive to London's character. The twenty years which might have been used in replanning on these lines have been wasted in desperate make-shifts and piecemeal 'improvements'; in the process many admirable features have been lost and immense sums of money expended but London is neither a finer nor more efficient capital city as a result.

THE PATTERN IN 1945

Within a short time of the end of the War a partial 'blue print' for post-war building had emerged. It was accepted that London and the war-damaged cities would need drastic redevelopment; it was accepted that this would entail decentralization of a large number of Londoners to new towns; it was less definite whether new towns would be needed for the other large cities.

Regional plans were prepared for the West Midlands, Clyde-side, South Wales and other regions but there was no national plan for the distribution of population and industry, no effective policy to check the drift to London and the South-East. The trends of industrial growth which mainly caused this drift were of expansion in the South and Midlands, relative stagnation in the North of England, Scotland, Wales and Northern Ireland. In these older industrial areas employment declined while jobs became more plentiful in the new industries such as light engineering and electrical equipment, in the South and Midlands.

Passing down the scale to the planning of individual towns the picture becomes somewhat clearer. Plans were prepared for many of them. Traffic circulation was usually on the principle of inner and outer ring roads connected by radial roads and the residential areas were disposed between them and described as 'neighbour-hoods'. The problems of central shopping and business areas were seldom examined in any great detail. The emphasis, rather naturally, was on housing; the immensity of the coming traffic problem was usually not foreseen, probably because very few new private cars had been on the market since the War. Indeed the Ministry of War Transport's *Design and Layout of Roads in Built-up Areas*, published in 1946, gave no hint of looking ahead to the days when production of cars would return even to the post-war level.

The neighbourhood unit was generally advocated as the universal panacea for all the earlier failings in the design of residential areas. It was not realized that this idea was not applicable, without modification, to the conditions of life in post-war Britain. And as Clarence Stein, one of its strongest supporters, has said 'The Neighborhood Idea is somewhat nebulous. It is difficult to separate the elements.'[1] Indeed, analysis shows that the approaches to it have two quite different sources. The early concepts had an openly sociological basis. It has also been claimed, particularly in later efforts to develop neighbourhoods, that they have a strong functional purpose, being a convenient way of subdividing towns and providing a pattern of social services.

The neighbourhood as a *social* unit was certainly the dominant approach when the concept was being formulated. Howard intended his wards to be almost self-contained units in themselves; the Barnetts aimed at bringing all classes together to live in helpful neighbourliness; Unwin's work was directed towards putting this into effect. The same purpose of fostering community life was clearly expressed by Clarence Perry in his original monograph on 'The Neighborhood': 'All cities contain residential areas which have local community characteristics in more or less degree. As civic or social entities they are, however, being continually destroyed or endangered by the spread of population, business or traffic. . . . Even if the metropolis tends to destroy the communities which it engulfs in the course of expansion it creates new ones by its own internal process.' Experience of life in big cities bears this out; it was perhaps particularly marked during the Blitz on London, when local loyalties and mutual aid, some

[1] C. Stein, *Toward New Towns for America*.

65

existing already, some dormant, and some arising entirely out of the conditions of the moment, grew and flourished.

It is hardly to be doubted that a community pattern existed over wide areas of Europe up to the time of the Industrial Revolution and in a weakened form survived its destructive effects. As Lewis Mumford has said '. . . the recently posed question of whether neighborhoods actually exist, particularly within great cities, seems a singularly academic one, indeed downright absurd in the suggestion that neighborhoods are wilful mental creations of romantic sociologists. Paris, for all its formal Cartesian units, is a city of neighborhoods, often with a well-defined architectural character as well as an identifiable social face. The Parisian neighborhood is not just a postal district or a political unit, but an historic growth; the sense of belonging to a particular arrondissement or quarter is just as strong in the shopkeeper, the bistro customer or the petty craftsman as the sense of being a Parisian.'[1]

Associated with this issue, in Britain, was the question of social mixture. Running right through the work of British pioneers of residential planning from Unwin onwards is a recognition of the evils of large scale segregation of classes into separate areas. This segregation was a legacy of the Industrial Revolution and the appalling conditions which led to the building of 'Housing for the Working Classes' by local authorities, and, after the First World War, the escape from the bye-law street to great estates of 'desirable residences' put up by speculative builders, for the lower middle classes. Decades of large-scale segregation had made it possible to 'place' a person socially by knowing where he lived.

The drabness of social life and the architectural monotony of vast numbers of similar houses became widely recognized, especially through the efforts of the N.C.C.S. and the work of eminent planners such as Abercrombie and Sharp. It became part of official doctrine to try to break down social tensions by mixing social classes—the working class and the middle class—

[1] L. Mumford, 'The Neighbourhood and the Neighbourhood Unit', *Town Planning Review*, January, 1954.

within the same neighbourhood unit. The degree of success attained is examined in a later chapter.

The other foundation on which the neighbourhood theory stands is functional. The argument is that the services of a town (schools, shopping, open spaces and so on) can most efficiently be provided if the town is regarded as a collection of smaller units. A corollary to this is the definition of units by through traffic roads and the exclusion of all but local traffic from residential areas. This is an attractive idea which breaks down the complex structure of great cities into units of comprehensible size. In practice, however, it has proved more difficult to decide on the right size of unit and even the considerations on which size should be based; this is indeed closely allied to the sociological problem of community feeling.

The traffic element in the neighbourhood theory–the aspect stressed by the Radburn plan and by Alker Tripp's 'precincts'– has never been disputed. It has just not been taken seriously. No proof is required that small children will not go more than about a quarter of a mile to a playground, that inconvenience and even hardship results from having to walk a long distance to school or shops, or that all these daily journeys are dangerous if a busy road has to be crossed. Yet those points were not sufficiently stressed after the War–and are not now. It was–and is–tacitly accepted that cars and tradesmen's vans should be able to draw up to the front door of every house, and that freedom of movement for vehicles should be the first consideration in designing residential areas. The massive demand of all families for a car was not foreseen. After a war in which mobility played a decisive part we were unprepared for mobility in times of peace.

CHAPTER 4

The Emerging Motor Age

It is curious that in Britain fifty years of the development of motor vehicles passed by before it began to be realized that here was a new form of transport which would completely upset our towns, creating a town planning revolution. Right up until the end of the 1940's, although there was certainly increasing concern about road congestion, the motor vehicle was not viewed as a problem with economic and social consequences of national dimensions. The private motor car, in particular, was still regarded as a middle-class status symbol, a means of private enjoyment for the more leisured and a means of door-to-door transport for the minority, giving expression to their elevated status in life. Even after the war, when large-scale production of private motor cars slowly picked up again, it was not until 1949 that the number of motor cars in use once more exceeded two million – the figure reached in 1939. By 1957 the number of motor cars in use had doubled – to more than four million; by 1964 there were eight million – four times the 1949 figure. With at least a million and a quarter registrations of new vehicles each year since 1959 it can be said that the 'motor age' had really arrived in Britain.

It is often inferred that the oncoming motor age could not have been foreseen. Had we known, the argument runs, we should of course have taken steps in time. But at least the new towns might have been expected to take up the challenge. As will be shown later[1] the first fourteen new towns were designed in a way which takes little more account of traffic than the earlier pattern. Yet the writing had been on the wall for a long time, certainly long enough, one would have thought, to be observed and noted by

[1] See Chapter 5.

men of discernment. The American example was before us even in the 1930's. As Clarence Stein has pointed out, even in the 1920's American cities had become nightmares of danger; in 1895 there were only five automobiles registered in the U.S.A.; by 1928 there were twenty-one million. Stein comments about the situation at the end of the 1920's: 'Pedestrians risked a dangerous motor street crossing twenty times a mile. The roadbed was the children's main play space. Every year there were more Americans killed or injured in automobile accidents than the total of American war casualties in any year. The checkerboard pattern made all streets equally inviting to through traffic. Quiet and peaceful repose disappeared along with safety. Porches faced bedlams of motor throughways with blocked traffic, honking horns, noxious gases. Parked cars, hard grey roads and garages replaced gardens.'[1]

The background of the emerging motor age in Britain in the 1920's and 1930's has to be more thoroughly understood if we are to learn the lessons from it. The menacing aspects in this period were not so much the overall increase in the numbers of motor vehicles but the consequences of a changeover to the use of motor vehicles instead of railways in a planning situation that was ill-equipped to deal with such consequences.

In particular the onrush of the motor vehicle brought about a powerful movement of population with radical consequences for town planning. The gradual swing-over to the transport of materials and finished products by road made changes in the location of industries possible. The old orientation around sources of coal and steel gave place to a new concentration around London and in the Midlands, especially for the new, lighter industries–engineering, electrical goods, vehicles, precision instruments, etc. The urban communities of the nineteenth century had gathered round the main lines of the railways and in certain cases, the canals. The increasing use of the motor vehicle in transporting goods by road was the greatest loosener of relationships between industry and the traditional areas of industrial concentration. This point has been stressed by Colin

[1] C. Stein, *Toward New Towns for America*.

Clark who says that: 'the railways immeasurably cheapened this long distance transport. But once the goods had been unloaded from the train or ship on to a horse-dray their costs of transport were very high. The inevitable consequence of this was that industry was concentrated in compact and densely populated industrial towns, or directly along the waterfront in sea ports. Motor transport unfroze this concentration. . . . The road transport age which followed the railway age has set the manufacturer and trader free from the close confines of industrial cities, within which they had previously been held.'[1]

The drift southwards of population from Scotland and the industrial North as well as from Wales and Northern Ireland towards Southern England and the Midlands has been a vital development influencing the whole national economy and the town planning problems of our urban populations. The motor vehicle, by freeing industrial locations, was the main factor which made such developments possible and gravely accentuated so many of the planning problems of Britain.

The other momentous change produced by the motor vehicle was to encourage the expansion of all cities and towns; the development of motor transport made it possible for town dwellers to live over a far wider area, cities and towns 'exploded'. True, the early developments of motorized public transport—particularly the tramcar and motor bus—had already started this process quite early in the century, but the private motor car greatly accelerated and extended it. Whereas the tramcar and motor bus were significant for extending residential development along radial roads, the motor car produced the infilling of the wedges of land between the main radial roads. Thus the dispersal of population from towns and cities increased, continuing as the faster-moving urban sprawl of the early 1930's and onwards. These 'explosions' spread rashes of sporadic building across the countryside and, in particular, the familiar ribbon development of housing along miles of roads leading into all our towns and cities (Plate 6).

[1] Colin Clark, 'Transport, Maker and Breaker of Cities', *Town Planning Review*, 1958.

Coupled with these two trends can be seen the beginnings, through the motor vehicle, of mass recreation. The countryside around the towns and cities soon became regarded not so much as agricultural land with an economic function but as legitimate play space for the townsfolk in their leisure periods. The effects of such mass recreation on the use of land were largely un-controlled. Commercialism was rampant everywhere; what planning powers existed at that time were not able to control even the worst excesses. The weekend bungalow, shack or 'chalet' invaded the countryside and especially the coastline. Today, if one contemplates the outcome of this avalanche of shanty town building, permanent caravan sites and rabid commercialism on some of Britain's seaside and country environments, it seems almost impossible that the mess created can ever be sorted out into a pleasant environment for enjoyment and relaxation.

Measuring the Problem

The growth in car ownership is a good indicator of the extent of traffic on the roads, the average annual mileage per individual motor vehicle having remained remarkably constant over the past decade. As a consequence, the increase in traffic on the roads is closely paralleled by the growth of car ownership–total vehicle mileage doubling between 1955 and 1965, while car ownership has also doubled.[1]

In spite of this large increase in traffic, it is not generally appreciated that paradoxically the private motorist is now better catered for in many respects than he was in the early 1950's. Average journey speeds have shown a remarkable constancy on many rural and indeed urban roads and in some cases have actually increased. Professor Smeed has shown that the journey speed of traffic on the main roads in Central London has not deteriorated since 1947.[2] In 1947 it was 11 m.p.h. and is about

[1] Ministry of Transport, *Highway Statistics, 1965*.
[2] R. J. Smeed, 'Traffic Problems in Towns', *Traffic Engineering and Control*, November, 1964.

11 m.p.h. at the present time. Yet, during this intervening period, the number of registered motor vehicles in the London area has almost trebled. Nor is this tendency unique for the metropolis. In both Glasgow and Slough, where studies were made, speeds in the early 1960's were actually higher than they were ten years previously.

The lot of the private motorist has improved in a further respect. Both the cost of buying and operating a car has been rising less quickly than the general level of prices. The costs of private motoring have been falling in *real* terms, a situation not equalled, unfortunately, in the public transport sector. On the other hand, the motorist has been faced by an increasingly difficult parking situation, a situation likely to worsen as parking meters and other parking restrictions are extended to still more towns and cities. Nevertheless, a survey in 1964 showed that 89 per cent of car commuters were still parking free of charge, most of them within 100 yards of their place of work.[1]

The 'motor age' is not, of course, a phenomena peculiar to Britain. In the United States, car ownership levels have always been much higher than in Britain and are still on the increase. Comparing different countries for their relative congestion proves to be extremely complex; the size of a country, distances between its towns and cities, the relative importance of public transport as well as average number of miles travelled by motor vehicles each year are all factors which make direct comparison almost impossible.

A crude measure of the increasing intensity of the problem in different countries (though not necessarily a guide to its relative seriousness in the different countries) is given by the changes taking place in the total number of vehicles and the total length of road in each country. The table[2] on the next page gives the picture in many European countries.

It will be seen from the table that whatever may be the relative position of many other European countries compared with Britain, their relative rates of vehicles per road mile have

---

[1] *People and the Motor Car*, Birmingham University, 1964.
[2] Derived from *Basic Road Statistics: 1966*. British Road Federation.

THE EMERGING MOTOR AGE

VEHICLES* PER ROAD MILE

	1954	1964	Percentage increase 1954–1964
Great Britain	22·3	49·7	123
Italy	9·8	43·4	343
West Germany	14·2	40·7	187
Sweden	11·1	30·8	177
Switzerland	9·3	29·6	218
Belgium	13·7	25·1	83
Denmark	7·8	24·3	212
Netherlands	13·2	23·9	81
France	8·2	19·5	138

* Vehicles in this table include cars, buses, coaches and goods vehicles only.

been increasing rapidly; they are entering a motor-age crisis of similar dimensions to that in Britain.

The problem is primarily one of traffic in towns; it lies in the very successes achieved in keeping the traffic moving. Towns as places to live in have been sacrificed for the traffic. Quiet residential streets have been turned into major one-way traffic arteries. A proliferation of street furniture–signs, signals, symbols, flashing lights, meters, white and yellow paint–these are the 'solution' being administered. Total road casualties in Great Britain in 1965 were almost 400,000, the majority of accidents taking place in urban areas. Increased traffic has meant increased pollution of the atmosphere from fumes. In squares and streets of the big cities there are reports of trees dying from the effects of motor fumes. In 1963, a government enquiry drew attention to the fact that in London, road traffic was the main source of annoyance from noise.[1] This, then is the problem–the harmful effect of the motor vehicle on urban environment.

And what does the future promise? With higher living densities than in the United States and perhaps a determination not

[1] *Report of the Committee on the Problem of Noise*, The Wilson Report, H.M.S.O. 1963.

to see public transport decline, our car ownership levels may not reach as high as the North American level. Be that as it may, the fact remains that the traffic situation is going to get a good deal worse.

If journey speeds are *not maintained*, then journey-to-work travel periods will lengthen with consequent economic losses to the nation. If journey speeds in our cities *are* maintained, then the danger is that this may be achieved at the expense of the quality of urban areas as places in which to live. Either way, the prospect is bleak.

COMMUTING AND PUBLIC TRANSPORT

The effects of the motor boom can only be contained within a reasonably civilized pattern of life if certain issues, not previously considered of very great significance, are squarely faced. By far the biggest single issue is that of commuter traffic. It is the travel at peak periods which causes the greatest problems, the greatest intensity of use of the road system and the greatest congestion. This raises two interrelated series of questions. Are we really satisfied that, in our new planning, we are disposing the elements –homes, factories, offices, schools, shops, etc.–in such a way as to reduce twice-daily, time-consuming, costly and wearing journeys to work? (Plate 7). Is there not great scope, even in our older towns and cities, for altering relationships between elements so that journeys to work are reduced and are undertaken by a smaller proportion of the population?

The other series of questions relates to working hours and leisure time. Is there no way of gradually adjusting our working lives, our shopping and leisure activities so that there is a far greater spreadover of times, particularly of starting and stopping these activities? Staggering of hours of work has long been advocated but comparatively little has been achieved by exhortations to offices and firms or by urging the jaded travellers themselves to 'avoid' travelling during rush-hours. If it is really so useful to stagger working hours, and there must be a few people who would doubt this, then surely it is necessary to seek effective measures to bring it about? Another aspect of spreadover of times concerns leisure time and holidays. We still tend to assume, perhaps because it has operated for so long, that we should all take our leisure more or less at the same time – on Saturdays and Sundays – and our holidays during July and August. Could not annual holidays range over a longer period by the comparatively simple device of staggering the summer holiday period of schools which, at present, tie so many family holidays to the peak period? Some adjustments to school examination arrangements might be needed but this could be done if it were justified. These measures could help to reduce congestion at peak periods and increase the efficient use of transport facilities. The strait-jacket to our thinking on such questions has to be tossed aside if such remedies are to make an effective contribution to tuning our lives to the motor age.

Closely linked to problems of commuting and travelling at peak periods is the future of public transport. Again the lessons from the U.S.A. are painfully clear; this country, in which the motor age has advanced furthest, has long neglected its public transport system. There are now growing regrets for this short-sightedness, and costly measures are having to be taken to revive public transport systems. Over forty United States cities are currently experimenting with new public transport facilities. In Britain, fortunately, after years during which the yardstick for public transport was economic viability – that public transport must 'make a profit' – there are signs of a change in attitudes.

Nevertheless it is still a common view, particularly among motoring enthusiasts, that public transport has no future. This view arises from a failure to appreciate that, without the radical redesign of our cities, it is virtually impossible to deal with commuter traffic in any other way than by mass transport. In New York, for example, in 1959, 75 per cent of those travelling into the city to work did so by some form of public transport. In London, over the same period, a million people travelled to the centre to work each day by public transport and only 70,000 by private car.

Forward-looking plans for the future assume that large-scale use of public transport will be essential; the plan for Cumbernauld New Town, which assumed car ownership of 0·7 cars per family for a compact, small town of about 70,000, assumed that there would need to be 42 per cent going to work by public transport and another 13 per cent walking. Fort Worth, a most ambitious and costly development project in the United States, with a complex, multi-level city centre, assumed that the number of commuters using public transport would have to rise from 17 per cent at present to 50 per cent.

Many changes will be needed in public transport. So far there has been very little really deep thinking about this. Not only new ideas and new forms of transport will have to be considered, but new combinations of public transport may be needed. Perhaps we should get rid of the terms 'public' and 'private' transport. Public transport has its innuendos of transport in the mass and in the raw; private transport has its relatively superior connotations conveying a symbol of status. Perhaps the term 'community transport' would give a more accurate indication of its social role without stigma, while the term 'personal transport' would perhaps indicate more fully the limitations of the personally owned motor car, especially for commuting.

THE MOTOR VEHICLE IN RESIDENTIAL AREAS

As the number of families with a motor car increased, so the inadequacies of the provision for the motor vehicle in residential areas has been more starkly revealed. In the older residential districts it has now become accepted that the public highway should be used for permanent private parking. If streets adjoin a shopping centre the carriageway often becomes a car park; the local resident has not even the facility of parking outside his own front door but must take his turn with the others and park as best he can. In newer residential districts, built since 1945, there was nearly always too little parking space and far too few garages to meet even present-day requirements. Even in the area of middle-class, larger detached and semi-detached houses, built by private builders or developers for sale, one garage per house is proving inadequate. Provision for the small car for the middle-class housewife, cars and motor cycles for older teenagers and parking for visitors and service vehicles have nearly always been insufficient. In much of the private housing built between the wars there was space for a garage either at the side of the house or at the rear, but few garages were built with the houses or considered as part of the original layout. The spaces have been filled up with garages provided by the house owners in every conceivable material, style and construction, resulting in a sharp deterioration in the appearance of the area, especially where the garages are part of a shanty town in miniature at the rear of gardens, mixed up with chicken-wire, potting sheds, greenhouses, compost heaps and old junk.

The greatest need for garaging and parking is not only in the older areas but also for housing estates owned by public authorities and built between the wars. Frequently these estates were built without provision for any garages, doubtless on the assumption that those who rented municipally-owned houses or flats would never be able to afford to have a car. Nor are deficiencies in the supply of garages on municipal housing estates confined to housing built pre-war; large areas of housing built since 1945 are grossly deficient in garages. This was revealed in a survey of the garage problem in Coventry.[1] At the time of the survey there were 1,540 garages in municipal estates but a waiting list of those wanting garages of 1,800, the majority being required on estates built since 1945.

There is also the problem of commercial vans and lorries being parked overnight and during weekends in residential areas. Such commercial vehicles aggravate the problems of parking and garaging, especially heavy lorries which are parked in the roads and parking spaces by residents who are lorry drivers by trade. This has now grown to constitute a major abuse of the public highways which it is most difficult to prevent without fresh legislation and enforcement action.

[1] R. W. G. Bryant and J. Mattocks, 'A Study of the Garage Problem in Coventry', *Journal of the Town Planning Institute*, June, 1959.

TRAFFIC AND ENVIRONMENT

The problems of the motor age are still regarded by most people as those of dealing with a very large increase of traffic on the roads. Yet we are really faced with a different environmental situation of crisis dimensions.

The problem can be simply stated. Assuming that it is impossible for all activities to be performed by men in motor vehicles, there must surely always be men on foot. Even the most inveterate drivers have to be pedestrians for a large part of the time. In urban conditions this produces a clash, which is becoming more and more difficult to resolve as the numbers of motor vehicles increases. Man as pedestrian—and 'man' in this sense includes a high proportion of women and children—must obviously take precedence over man as motorist. This may not always be readily accepted; give a person a vehicle with lethal power and he often seems to cease to have a human attitude towards his fellow creatures—but we all have to be and indeed want to be walkers for a substantial part of our lives. Many people might want to walk even more if some of the present dangers and unpleasantness of moving about in our towns and cities could be reduced.

This approach led Colin Buchanan to put forward his important concept of 'environmental areas' or 'environmental units' whereby towns became subdivided into areas where vehicles only have very limited penetration and the pedestrian is dominant (Figs. 12, 13). These environmental areas are to be connected to the rest of the town by a network of roads for distributing traffic. This approach has aims which are similar to the 'precincts' of Sir Alker Tripp;[1] but it does not presuppose that such areas or precincts should be only for one use—industrial, commercial, residential, etc.—nor does it assume that roads and pedestrian areas have to be at ground level; separation of vehicles and pedestrians on different levels adds a third dimension to the precinct theory: 'It might be objected that this is no more than the old precinct theory brought up again. In a sense it is, but it is

[1] See Chapter 3.

in a simpler and basic form. I go no further than to postulate areas where the pedestrian environment is paramount, linked by a distributory network for the traffic, whereas the precinct theory was always tied to the idea of areas of "homogeneous use" bounded in the same horizontal plane by traffic arteries. On my

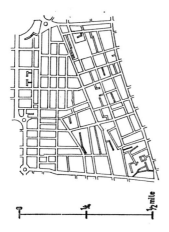

Figure 12. The existing road system within towns. Traffic penetrates everywhere.

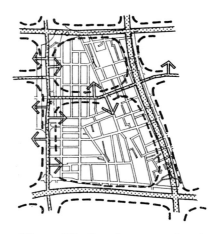

Figure 13. Creating an ordered, distributing network, converting existing roads or cutting through with new ones. Between main roads are created 'environmental areas' with limited access for traffic.

definition the distributory network could as well pass underneath the environmental unit as round it.'[1]

Environmental areas are related both to Alker Tripp's precincts and also to the neighbourhoods of the County of London Plan. This is acknowledged in the Buchanan Report: '. . . Sir Alker Tripp was advocating something on these lines over twenty years ago, and the precincts and neighbourhoods of the County of London reflected the same approach. But in the

[1] Colin D. Buchanan, 'Standards and Values in Motor Age Towns', *Journal of the Town Planning Institute*, December, 1961.

face of the rapidly increasing number of vehicles it acquires a new urgency; it now requires to be explored and developed from a mere concept into a set of working rules for practical application.'[1]

This concept of environmental areas has attracted the support of many town planners, architects and engineers. It deals in simple and fundamental terms with the essence of the problem of the motor age. It re-establishes the primacy of the scale of values of man on foot; at the same time it calls for changing social conditions and environment to suit the new needs of the motor age–to utilize the advantages of greater mobility without surrendering the mastery to the machine.

Related to this concept of environmental areas is the possibility of re-examining the elements of a town, the disposition of its functional parts–the very stuff which determines the number and the distances of the journeys done. It has been said by Professor Buchanan that motor traffic can be subdivided into two categories–'purposeful' and 'optional'. Purposeful traffic includes business, commercial and industrial traffic; this traffic, it is argued, cannot be reduced without endangering the very fabric of our economic life. The 'optional' category includes commuting by car rather than public transport and personal movement about the town, journeys which could be taken in some other way than by car or journeys simply for pleasure. If anything has to go, it is said, it is the 'optional' journeys rather than the 'purposeful'. This seems much too static a way of categorizing traffic movement. For it is self-evident that in some cases journeys which might be classed as optional (commuting by car because public transport is either inadequate or very unpleasant) are really almost unavoidable. On the other hand many journeys which in the above categories would be classed as purposeful need not remain so if big changes in the organization of commerce and industry occur. For example, in some circumstances, it has been claimed, operational research techniques could save up to 20 per cent of commercial vehicle mileage.

[1] *Traffic in Towns*: The Buchanan Report, Ministry of Transport, H.M.S.O. 1964.

4. The Royal Crescent, Bath: urbanity through order.

5. Charlotte Square, Edinburgh: fine architecture and open space
near the centre of a capital city.

6. Ribbon development along main roads; the legacy of the
inter-war years.

7. Traffic congestion at peak periods.

The reduction of many of our present purposeful journeys may also be produced more by technological advance than by town planning. For example, the not-too-distant possibility of the video-telephone—where one is able to see people as one talks to them—could make a considerable difference to the amount and frequency of personal visiting necessary in the commercial and industrial world. It might become a great deal more common for many business people to work from their homes and to reduce the number of days they go to 'the office'. If this were coupled with the moving of much routine office work into closer association with residential areas this could help to produce a profound change in the journey-to-work and the rush-hour problems.

The idea of environmental areas does not mean a rigid subdivision of areas into separate zones of uses for different activities—residential, factory, shopping, office, etc. The sharp separation of uses in our towns is the main factor generating vast amounts of unnecessary movement about the town—from home to work, to shops, to entertainment, etc. In particular, the relation between work—mainly factory or office—and home is the one crying out for radical changes which could reduce the amount of purposeful journeys. Some industry is obviously highly unsuitable for association with residential areas—because of noise, smell, fumes, heavy traffic, etc. (although even these aspects are more controllable than is often thought if industrialists are prepared to take the trouble and allow the extra expense). But many light industries, especially those employing women workers part-time, are quite suitable for relatively close association with housing in a residential area if care is taken in planning. The trouble with most of the larger, planned residential areas that have been built in new towns and in town redevelopment or expansion schemes since 1945 is that they are *too residential*. The introduction of suitable light industry, offices and other uses will reduce the amount of commuter traffic as well as providing some welcome architectural contrast and interest in residential areas.[1] Care would have to be taken, of course, to avoid industrial and

[1] For full treatment of this question, see Anthony Goss, *British Industry and Town Planning*, Fountain Press, 1962.

commercial traffic going through residential areas; the siting of small factories and offices would therefore tend to be on the periphery of environmental areas. But it would be possible for a much larger proportion of people to walk to their work and the 'peak hour' traffic burdens would thus be very much lighter.

There is perhaps a tendency in our society at present to pride ourselves too much on our mobility without examining carefully

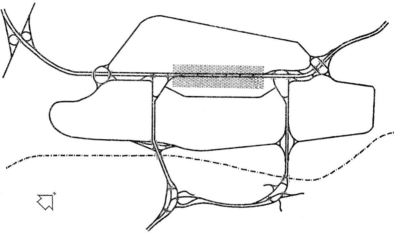

Figure 14. The road layout of Cumbernauld New Town, designed to cater for seven out of every ten families having a car. All main junctions need to be multi-level intersections.

enough how much of the mobility is just a waste of time and energy. Many people undoubtedly often derive pleasure from driving or being driven in a motor vehicle. But in present-day traffic conditions, inevitably worsening in the future with the vast increase of cars on the roads, most people will only move about in motor vehicles because they want to go somewhere rather than because they enjoy the sensation. The romantic age of the joys of motoring will largely belong to the past as far as urban motoring is concerned. In helping to plan our society, town planning must enable mobility to play its full part but without allowing the balance to tilt away from the prime aim—the creation of a well-ordered human environment.

82

THE EMERGING MOTOR AGE

THE FUTURE OF THE MOTOR VEHICLE

The 1950's saw a gradual awakening to the fact that the motor age had arrived and increased trepidation about the problem it was creating. A small band of experts, especially Colin Buchanan, the Urban Road Planning Adviser to the Ministry of Transport, sounded alarms with vigour and effect. This was undoubtedly necessary because of the propensity which the Ministry of Transport itself had shown, time after time, to under-estimate the increase in traffic to be expected. In 1945 the Ministry estimated traffic increases by 1965 would be 75 per cent over 1933 figures in urban areas and 45 per cent more in rural areas; but by 1950— five years later—the number of vehicles had already nearly doubled compared with 1933. In 1954, the Ministry told highway authorities to plan for a 75 per cent increase in traffic by 1974. But by 1962 the number of vehicles was already 75 per cent above the 1954 figure. In 1957, the Ministry forecast that there would be eight million vehicles by 1960; but there were nine million. In 1959, the Parliamentary Secretary to the Ministry of Transport forecast that there would be twelve and a half million vehicles by 1969; but this figure was passed before the end of 1964.

But even with correct forecasting of need, the money has still to be found to build the roads, to say nothing of the large-scale renewal in urban areas, especially in their centres, which should go hand-in-hand with new roads. In 1964 about £181 millions was estimated to have been spent by Central and Local Government on new construction and major improvement of roads; this was, of course, more than the £20$\frac{1}{2}$ million spent in 1939 but not very much more bearing in mind the almost fourfold increase in prices.

Nevertheless, since 1957, road expenditure has been taken more seriously by the Government. There has been a reluctance to cut-back spending on roads at times of general economy in Government spending; there are signs that the rate of increase of spending on roads is now greater than the rate of increase in traffic. But there is an enormous backlog to catch up; £181 million does not go very far when urban roads in central areas

83

cost £5–10 million per mile or new dual carriageway roads in an urban area cost about half a million pounds per mile.

In thinking about the future it is not only necessary to have some idea of the size of the problem, but also to know more in much greater detail about the way traffic moves about at present to help in the design of road networks which will cater for much larger volumes of traffic in future. The techniques for achieving this are being developed in the form of transportation studies, traffic engineering and traffic management. But these techniques are in their infancy. Most of the road schemes that have already been embarked upon in our towns and cities were undertaken before most of the fundamental research about the complexities of traffic movement had been done. They had to rely on guess-work and intuition to a large extent. The wonder is that they serve as well as they do though further increases in traffic are already revealing their inadequacies.

A great deal more traffic survey work needs to be done but even then the information obtained has to be used with great care because it can be very misleading. Often the 'need' seems to be established by traffic surveys for existing roads to be far wider and to have more complicated junctions in order to take the increased flows of traffic. As Alan Day, the economist, has said: 'The fact is that traffic surveys can say very little about "frustrated" traffic, and also they can say very little about future changes as people change their place of work and residence. Nor can they show what will be the effect of new traffic facilities, both on work places and where people will live. The answers to these questions need intelligent thought as well as mere observation of what is happening now.'[1]

Survey of present patterns of traffic movement provide valuable information which can be an aid to planning. But, as with many forms of survey work and data collection needed for town planning, the value of the information obtained depends on the relevance of the questions asked. Evaluation is still required and this must be related to other factors such as changing patterns of land use, changing intensities of traffic generation by different

[1] Alan Day, *Roads*, Mayflower Books, 1963.

uses as well as the all-important policy question of the relative importance of mass transport and private vehicles. Traffic surveys are no substitute for skilful town planning.

It is frequently argued in the rail versus road controversy that rail must needs go to the wall because 'the railways do not pay' – i.e. do not make a profit. There is, unfortunately, no appropriate balance sheet available which gathers together all the legitimate expenditure on roads. But the data that is available seems to show fairly conclusively that our roads are also a heavy liability to the taxpayer even though most of this liability is hidden. The roads can claim little in the way of income except by motor vehicle and fuel taxation. But what do the roads cost to administer, police, maintain, and what are the consequent costs of congestion and the heavy accident burden of the roads? The biggest item is probably the cost of congestion, not only wasteful expenditure on motoring because of uneconomic speeds, but more particularly the appalling losses in time and human energy which must eat great holes into our national productivity. It is not easy to gauge these losses by congestion but it is certain that, despite the enormous sums that will be spent on roads to minimize such losses, they will still increase. The Road Research Laboratory has put the present cost of traffic congestion at a figure of at least £200 million a year – or £500 million if allowance is made for time lost in traffic delays to people going to and from their work; and this figure, they say, is increasing every year.

The cost of accidents is also far higher than most people realize. In 1965 the cost of road accidents to the community was estimated, by the Royal Society for Prevention of Accidents, as £264 million, including hospital treatment, loss of earnings, damage and repair to property, compensation for personal injury and administration costs; this is over £650 per accident. But who is to put a value per head on the dead and maimed, the death roll varying between 5,000 and 7,000 a year, and the injured varying between 250,000 and 350,000 a year for the past ten years? And who is to say what the figures of dead and maimed will be with double or treble the traffic on the roads? Yet we have come to accept as 'normal' an annual toll of bloodshed on the roads that reads like casualty lists from a major war.

Road transport now has a vital relationship with town planning. There can be no effective town planning in urban areas without the will and the means to tackle boldly the problems of the motor age in Britain which, by all the evidence, is only just beginning. This was the powerful and urgent message of the Buchanan Report, *Traffic in Towns*.

But town planning will become a mockery if it is regarded as merely an exercise in catering for many more cars, a means of keeping the cars moving. There is a trend, and it is a powerful one, which would reduce town planning to a department of traffic and highway engineering, concerned with filling in the spaces between great new urban expressways and motorways of fantastic complexity. The motor age is the greatest challenge that town planning has yet faced. There is a powerful tendency for the machine and its adherents to sweep aside all recognizable human values and substitute the natural law of the asphalt 'jungle'. This is the spectre which will haunt all efforts to shape out towns to the twentieth century in the years ahead. This is the challenge of the motor age.

CHAPTER 5

Lessons from 'New Towns'

In Britain, 'New Towns' are no longer a theory; they have crystallized into large-scale practice and can be judged on performance. In all, twenty-one new towns had been started between 1947 and 1967 and fourteen of them are well on the way to completion. Their total population will ultimately reach more than one and a half million people. In addition to the *new* towns, developed through the machinery of the New Towns Act of 1946, there are now an increasing number of major town expansions taking place. Many of these schemes are of the same large-scale dimensions as the new towns; so that town expansion and new town building has been linked together in this chapter.

The bolder planning that was needed for new towns has been the subject of much uncritical adulation as well as critical comment. New towns were started in the late 1940's in Britain without an established background of recent planning practice of anything like such dimensions. There have still not been many serious attempts to assess their achievements and failings objectively or to learn from them. Too much of the criticism of the new towns has been negative in character or out of context—comparing them with other architectural or town planning schemes or projects of a different scale, with different costs or for different social and economic conditions.

Any serious discussion of the experiences of creating such large-scale urban units as new or expanded towns must start from a critical standpoint; there was so little to go on from previous experience–not only in Britain but from the rest of the world–that their planning must surely be regarded as experimental. But in seeking to draw lessons, we must be aware of several pitfalls. We are lumping together a considerable number of town-sized projects which have many differences—of size, shape, composition, economic base and of relationship with other urban units.

The first fifteen new towns started in Britain differed considerably from one another in their initial size and ultimate planned populations. Some, like Hemel Hempstead and Basildon, had already over 20,000 population within the area of the new town to start with. Others, like Harlow, Stevenage and Cumbernauld, were only villages in the new town area. The new towns of Newton Aycliffe and Peterlee were originally only going to be of about 20,000 population each; whereas many of the others were to be 50,000 or 60,000 population and Basildon was to have been at least 80,000 population. More recent town expansion schemes and new town schemes are of much larger dimensions. And the original new towns are nearly all, to different degrees, being expanded beyond their original target populations. The expansion of Northampton is to be from 120,000 to 222,000 people. Current thinking is for new towns or new cities with ultimate populations of 150,000 to 250,000.

The functions of these new and expanded town developments also differed considerably. In the case of the ring of new towns around London (Hemel Hempstead, Stevenage, Hatfield, Welwyn, Harlow, Basildon, Crawley, Bracknall) their links with London remained and substantial commuting to work into London has never been absent. Whereas Corby was really a town expansion scheme for a town dominated by one industry—steel. The degree of relationship with the cities and conurbations has also varied with, in some cases, a close tie between the new town and the nearby city that fostered it—as in the case of Cumbernauld and Glasgow. When the large-scale town expansion schemes now being advanced are also taken into account, the differences, particularly of function, increase.

Size and the Local Economy

The first series of new towns in Britain—the first fourteen—are now well on the way to the completion of their main constructional phase. In terms of attracting industry and population, there are no longer any doubts about their success. Office employment has proved to be more difficult although Crawley, one of the earliest, now has over 3,000 office workers compared with 17,000 employees in manufacturing industry. The town centres, although tending to lag behind the rapid additions of people coming into the towns, are now all flourishing and attract shoppers from a much wider area than the new towns themselves.

It has proved to be very difficult to produce planned town centres with real urban quality; a large number of developers and their architects, together with the proliferation of new materials and finishes and the strident clamour of advertisements (often behind the shop windows or from upper storeroom windows) has produced an exuberant but restless and cluttered appearance.

The temptation to profit further from success by increasing the size of the new towns before they are completed has proved irresistible. So that all of the first fourteen new towns will go on to become larger towns than was at first envisaged when their master plans were drawn up. In part this is sensible. The crude formulae, of 60,000 population and six neighbourhood units containing 10,000 each, has had to give place to more profound and subtle relationships of the whole to its parts. Economic factors have also shown, by experience, that there are substantial economies produced by building larger new towns. But the further expansion of the original fourteen new towns has led to difficulties. In many cases, the original master plans were conceived statically—in terms of a fixed optimum size which allowed only for a limited natural increase of the population after the end of the main constructional phase of building the town; the plans of many of these earlier new towns are therefore not easily adaptable to large-scale additional expansion. Putting it another way, had it been postulated at the beginning to those preparing the master plans, that such an additional expansion might be an eventuality that they must consider, it is certain that many of these plans would have been prepared quite differently; there comes a point where further accretions to a statically conceived plan endanger the efficacy of the original plan. Rather than additions to the edges of these earlier new towns it might be better if the new towns themselves were used as nuclei to a cluster of towns or villages suitable for expansion, giving to the new towns that wider context of sub-regional development which was largely lacking when they were first planned.

There may, however, be important exceptions to the undesirability of radical expansion of the earlier new towns and the case of Corby seems to be one of them.[1] It is a special case because its urban growth has come from the steel industry and the overwhelming preponderance of one employing firm–Stewarts and Lloyds Limited, who built their steelworks at Corby in 1934 near the ironstone beds, when Corby had a population of about 1,500. By 1950 the town's population had reached 15,700 with 5,000 of the town's workpeople employed at Stewarts and Lloyds–about 80 per cent of the labour force. The new town started in 1950 and was to have grown to 40,000 population. The proportion of the working population engaged in iron and steel was to have dropped to 55 per cent by the introduction of alternative industry. But, all the time the steelworks continued to expand so that, even though there has been an additional twenty-one factories and two industrial estates, employing an additional 3,500 workers (mainly women), the steelworks had 13,000 employees by 1966 compared with 8,000 at the commencement of the new town. So that about 60 per cent of the working population of the town are now employed at the steelworks, with the prospect of a further 2,500 employees by 1968. The new town of Corby, as planned at the beginning of the 1950's, now nears completion but its overwhelming specialism in employment in one industry–and, indeed, of one firm–is still its main feature.

Now a further area has been added to the size of the new town which will extend it by over a third more than its previous size. A second master plan was prepared in 1966 aiming to increase the size of the town to 75,000 by 1986, moving up to a possible 83,000 population by the end of the century. The proposal for

[1] See D. C. D. Pocock, 'Town Size and Diversification', *Town and Country Planning*, January, 1965.

further expansion of Corby is clearly an opportunity to help correct weaknesses in the original plan. Had there been a bolder concept originally there might now be a healthier position. Experience has shown that it has proved difficult to attract other male-employing industry because of the domination of the steelworks over the labour market. The small-scale of the town centre has also been a factor which has slowed the town's development.

The lesson here is the importance of a close study of the local economic structure and its place within a sub-regional economy before attempting to determine a town's size. A study made of another steel town – Scunthorpe – with a similar concentration of employment in the one industry concluded that, taking the potential for growth inherent in the physical structure of the town, together with the very high degree of specialization in the town's economy, the case for large-scale expansion was overwhelming. The present population of Scunthorpe is about 70,000 with 50 per cent of the working population employed in the iron and steel industry. It was concluded that: 'It seems doubtful if anything less than a doubling or even a trebling of the present population of the Borough can overcome the inherent difficulties of the town's high degree of specialization. Whether the expansion is seen as large-scale town expansion, for which the town is suitable, or as a part of a larger linkage of towns within a Humberside urban complex, depends on major national and regional considerations. But Scunthorpe's future depends on large-scale growth and inter-linkage with a wider regional pattern.'[1]

NEIGHBOURHOOD UNIT IN PRACTICE

In the early guidance given, at the beginning of the New Towns programme, by Government Advisory Committees and the official New Towns Committee itself, the solution to the problem of the subdivision of the town was seen as the neighbourhood

[1] See *Scunthorpe: A Study in Potential Growth*, Planning Research Unit, Leeds School of Town Planning, published by Scunthorpe Corporation, 1966.

unit. The neighbourhood units built in the new towns tend to be much purer in form than those adopted for suburban expansion of older towns and cities; less compromise was necessary to relate them to existing conditions. The neighbourhood unit in the new towns is therefore worthy of special examination, for this issue of subdivision of towns and cities has become so muddled by both adherents and opponents that it is necessary to try to seek a more objective judgement by viewing both what was intended and the actual outcome of these intentions.[1]

The main proposition on which neighbourhood units were based was one of size of population—as near to 10,000 people as possible. Whilst it was acknowledged that 10,000 people would provide more children than could be catered for by one primary school, it was assumed that it would be satisfactory for a neighbourhood unit to have two primary schools. Thus the basis of the primary school as the centre of community activity, the keynote of the neighbourhood unit as conceived by Clarence Perry, became impossible and the centre became, more often than not, a parade of shops with at least one of the primary schools sited a long way away. Thus the neighbourhood centre as a climax of social significance became emasculated; intimate and neighbourly closes of houses often led to an anti-climax—a few shops and a miserable timber sectional hut which was intended to serve, for at least a generation, as 'the community centre'.

In practice, the neighbourhood unit with 10,000 population proved unsatisfactory. Units as large as this tend to be poly-centred (see Fig. 15). Where low densities are adopted they also tend to be unsatisfactory in size—neither small enough for a sufficiently intimate community unit nor large enough for a district of the town. The amount of local shopping required, with self-service and supermarket retailing now firmly established, tends to be far less than was envisaged. The recommendation that neighbourhood unit boundaries should conform to ward boundaries has hardly ever been followed. The diminution of the neighbourhood centre and the difficulty of putting primary

[1] See Anthony Goss, 'Neighbourhood Units in British New Towns', *Town Planning Review*, April, 1961.

schools at the centre resulted in a relatively weak consciousness among new town dwellers of belonging to a sub-unit of the town.

One or two new towns adopted a variation of the neighbourhood unit–the 'Neighbourhood Cluster' for the subdivision of residential areas. This concept was used as the basis for the Master Plan at Harlow New Town by its architect-planner, Frederick Gibberd (see Fig. 15). As originally designed, the neighbourhood units at Harlow were comparatively small, designed to serve only 3,000 to 4,000 persons. There are four neighbourhood clusters at Harlow grouped around three major sub-centres, with the town centre serving as the centre for the fourth cluster. Thus each 'neighbourhood unit' has its own primary school and a few local shops but the 'neighbourhood centre' is at the junction of the neighbourhood units and serves a much larger number of people–varying from 17,000 to 25,000. The neighbourhood cluster concept, with smaller units based on the population needed to support a primary school and with main shopping at the meeting point of several neighbourhoods, is much nearer to the original concept of Clarence Perry and tends to be more successful. It provides smaller and more comprehensible units, centring the local community on a school rather than a shopping centre. The larger 'district centres' for shopping at the meeting point of neighbourhoods conform more to the changing pattern of shopping. It also overcomes the criticisms often made of new town neighbourhood shopping centres that the limited number of shops prevents adequate competition between shops and thus makes for higher prices.

The residential units proposed in the ill-fated project for Hook New Town also utilized a similar kind of residential structure, although taking much greater care to cater for the motor age. The residential structure was clearly related once more to primary school size, with other functions and social facilities placed in different residential areas without any attempt at arriving at a universal formula, but with pedestrian access to them from other residential units by placing facilities along the pedestrian ways (Fig. 16). As the study of Hook stated: 'The possible size-range for the superblock residential units at Hook would be determined

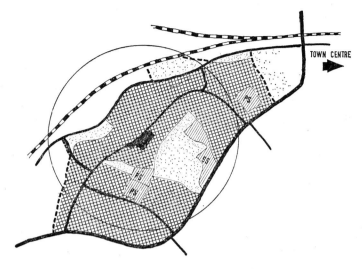

Broadwater, Stevenage (9,650 people)

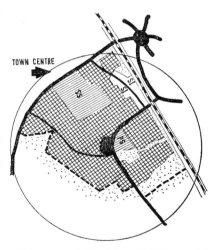

Kingswood, Basildon (6,000 people)

Figure 15. Comparison of three neighbourhoods in the New Towns, all drawn to the same scale.

Great Parndon, Harlow. A neighbourhood cluster for 25,000 people

CIRCLES ½ MILE RADIUS FROM
NEIGHBOURHOOD CENTRE

RESIDENTIAL

PRIMARY SCHOOLS

SECONDARY SCHOOLS

PUBLIC OPEN SPACES

NEIGHBOURHOOD CENTRE

SERVICE INDUSTRY

BOUNDARY OF UNIT

95

by a balancing or reconciliation of several factors–statistical, social and visual–varying with the housing density and the household structure from one part of the town to another. Within the inner town the ten-minute, half-mile, East–West walking distance . . . governs the length of the block. If the primary school is

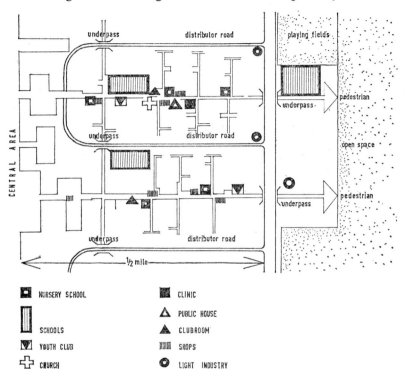

Figure 16. The diagram of residential areas from the study for Hook New Town. Separation of pedestrian and vehicle is proposed with social facilities grouped along pedestrian ways.

centrally located, this half-mile maximum length fits in with the child's quarter-mile maximum walk to school. With the inner town household structure, a two-form-entry primary school is needed for every 4,000 to 5,000 people.'[1]

The emphasis here is changed from 'neighbourhood units'

[1] *The Planning of a New Town*, London County Council, 1961.

each with an identical structure towards residential areas which vary in their composition. As with Cumbernauld New Town, the residential areas nearer the town centre are not given specifically neighbourhood centres, but are oriented towards the town centre itself. Again, as with Cumbernauld, beyond the 'inner town' there were to be separated neighbourhoods, with their own local, social and shopping centres; Cumbernauld has called these separated communities 'villages' but their degree of separation from the town as a whole is really not sufficient to justify this; despite the name, they are really neighbourhoods. Lewis Mumford has commented on Cumbernauld's community structure: 'Cumbernauld is very interesting because it tries to break away from the two phases of the original new town. It attempts to get away from the notion of the neighbourhood unit by not having differentiated neighbourhood units and by not having any spatial separation between the neighbourhood units—as in so many of the new towns. I don't think it is able to abolish the neighbourhoods; the neighbourhood will still reappear in Cumbernauld. On the other hand they've made a very radical and I think valuable departure . . . Cumbernauld is based on the pedestrian scale; every part of it, even the town centre, is within walking distance, half-a-mile or so, of the furthest residential neighbourhood . . . the gain in social life—the gain particularly from the standpoint of the neighbourhood—is a great one.'[1]

What seems to be emerging is the more flexible application of basically sound principles—the need for some form of subdivision of urban tissue. It is not just a question of the grouping of facilities and services to make it easier for practical planning. It is also a question of the social significance of grouping them together so that different activities gain in convenience as well as enhance opportunities for social contact by the proximity of one social facility to another.

RESPONSE TO THE MOTOR AGE

The response of the early new towns to the motor age can only be described as disappointing, particularly in their provision, in

[1] Lewis Mumford, *Journal of the Town Planning Institute*, November, 1961.

residential areas, for the garaging and parking of the motor
vehicle. These defects were not always the fault of the town
planners, architects and engineers who planned and built the first
fourteen new towns; many of them strove nobly to provide an
up-to-date road system separating pedestrians and motor vehicles
as much as possible and to increase the garage provision for
future needs; but many of the efforts were rebuffed by a 'penny-
wise' Treasury via the Ministry of Housing and Local Government. In a frank appraisal of one of the new towns, the former
chairman of its Development Corporation has said: 'If the Hemel
Hempstead Development Corporation and the Ministry of
Transport had known that the motor car age was upon them,
they would have produced "a very different sort of town".' One
of the things he would wish to do differently was to provide
more garaging: 'For the first six or seven years the Ministry
would not approve any layout which showed more than one
garage to every four houses. This was "totally, hopelessly
wrong".'[1]

In the decade whilst the first fourteen new towns were getting
under way, no additional ones were started. During this time the
number of motor vehicles was rapidly increasing in Britain; the
effects of this increase were belatedly being recognized. With the
planning of the fifteenth new town, Cumbernauld, which started
in 1956, a realistic approach to the motor vehicle was incorporated for the first time in a British new town. Here, in planning
for a relatively small town of an ultimate population of 70,000
people, the implications of large-scale car ownership were
examined. It was assumed that there would ultimately be 70 per
cent of all families in the town who would have a car, that 63 per
cent of the cars in the town would be in use together at the peak
hours and that 45 per cent of the population would travel to work
by car (17 per cent as passengers), while 42 per cent would be
travelling to work by community transport; 4,000 people, it was
estimated, would still be near enough to walk to and from their
work place. It was also estimated that 20 per cent of the working

[1] Henry Wells, former Chairman of the Hemel Hempstead Development
Corporation, *The Guardian*, 19th March, 1963.

population–7,000 people–would commute daily, mainly to nearby Glasgow.

When these estimates were translated into a road system it was found that only a radical approach would cater for road traffic of these volumes (Fig. 14). Two important principles were established. First, it was necessary to clarify the difference in function of various roads–the *development roads* serving the dwellings, the *distributor roads* which collect and distribute traffic from the development roads and, in turn, feed into the *main town roads*

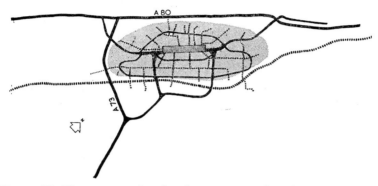

Figure 17. The system of pedestrian routes and underpasses linking the town centre and all residential areas at Cumbernauld New Town.

leading to *trunk roads* linking the town with the wider region and other towns and cities. The second principle was that foot and motor traffic should be separated to the maximum extent possible. This meant providing two separate circulation systems–for pedestrians and vehicles (Fig. 17). Only in exceptional cases were footways provided along main roads. There has to be a completely separate system of pedestrian ways and as far as possible it was intended that, whenever a pedestrian way crossed a road, an underpass or overpass should be provided; bus stops were to be located at interchange points between footways and roads.

Thus two separate and distinct patterns for movement about a town were envisaged; the pedestrian paths, as direct as possible (taking into account that people on foot like to be able to 'cut the corners), and the roads for motor vehicles. The pattern which

99

emerged was of roads leading motor vehicles outwards and around the edges of the residential areas and footpaths, taking people as directly as possible from home towards the centre of the town.

Cumbernauld can claim to be the first town in Britain where the road system was designed from the beginning on scientific principles to cater specifically for the motor age. It showed that a hierarchy of roads, performing different functions, was an essential principle to be followed by virtually every new town and for large-scale urban renewal. The main town roads were really urban motorways which served to take traffic from one part of the town and lead it to the national trunk road system. There were to be no junctions to these main town roads except at intersection points which were relatively few in number. In urban conditions it had been recognized for some time that it was the capacity of junctions to take traffic which determined the maximum flow of vehicles possible on any road system. Cumbernauld showed that even for a small town of 70,000 people, and even with comparatively modest assumptions – 45 per cent travelling to work by car – all junctions with the main town roads would need to be multi-level (see Fig. 14); roundabouts, which serve a useful purpose for much smaller densities of traffic, were no longer a feasible solution. This was a conclusion which has had a powerful and lasting effect on town planners as well as on many road engineers. The complexity of multi-level junctions in the United States had often been looked upon as either an extravagance, which was only possible where land and money were abundant, or as a fantasy in a world where fantasies abound. Cumbernauld showed that even in the ideal conditions of new towns, if the motor vehicle was to be adequately dealt with, multi-level traffic junctions were inevitable and so was a complex road system which few would have countenanced twenty years ago.

The new towns also demonstrated that fresh ways of dealing with motor vehicles in the centres of towns were needed.[1] The main break-through had already been made at Stevenage with

[1] See Chapter 7.

8. An interchange on a Californian freeway carving its way through a residential district.

9. Entrance to the city from the San Francisco Freeway.

10. Early housing at Harlow New Town. The skilful use of existing landscaping and new planting gives a semi-rural atmosphere in the best traditions of the 'garden city' movement.

11. A quiet cul-de-sac at Harlow. But the constant presence of parked cars is testimony to the inadequate provision for the motor vehicle.

the creation of a large pedestrian shopping precinct and large-scale car parking, sited around the edges of the shopping centre. The inspiration for this was undoubtedly the earlier Lijnbaan shopping centre at Rotterdam (Plate 39). The pedestrian shopping precinct has now firmly established itself in the town planning vocabulary so that no central area shopping scheme is regarded as complete without one; today it is strange to look back on the pugnacious opposition of the shopkeepers to this early pioneer scheme at Stevenage. Now, multi-level town centres are beginning to be accepted as more logical solutions for the motor age. Such projects, which would have been considered fantastic and costly dreams a mere fifteen years ago, are now being put forward and accepted. The town centres for Cumbernauld New Town, the proposals made in the studies for Hook New Town and the large town expansion schemes for Basingstoke and Andover in Hampshire all utilize some form of multi-level town centre as also do the more recent town centres for later new towns. That there is such wide acceptance of these bold solutions is a sign of the speed of change in our times.

HOUSING LAYOUT AND HOUSE DESIGNS

The new towns are also of great significance in the lessons which they produce about the design of housing on a very large scale. The early new town housing layouts were relatively conventional in pattern. The longer terraces adopted gave greater possibilities of effective architectural treatment of the front elevations compared with suburban semi-detached houses but the problems of the backs of houses and the lack of privacy of the back gardens remained largely untackled. Many of the new towns, built in a rural landscape setting, were outstanding in the efforts made to retain mature trees and shrubs. This care for existing landscape, more generous allowances for new planting, and the developing skill of landscape architects all helped to produce a pleasant, semi-rural effect reminiscent of the best arcadian suburbs like Welwyn Garden City, Hampstead Garden Suburb and Bournville. At first, the garden-suburb influence was

101

all-pervasive. However, as the densities increased, the arcadian idyll faded.

Earlier new town schemes were but poorly endowed with garages and parking. Frequently garages were provided for only a small proportion of the dwellings–perhaps a mere eight out of

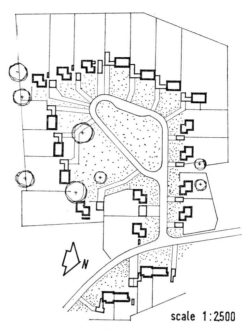

scale 1:2500

Figure 18. A group of related, detached houses at Basildon New Town. Single and double garages have been used to help link the group together (Plate 17).

every hundred dwellings at the most. As cars became more numerous, and vans and lorries, taken home by their drivers, were left outside their houses for the night or the weekend, the residential streets and culs-de-sac gradually became littered with vehicles. The response to this has been to step up the programme of garage and parking provisions, although the rents charged for garages have acted as a curb on the desire to have a garage, especially for the older car. Other methods of car storage have also been investigated, especially grouped car ports providing partial shelter.

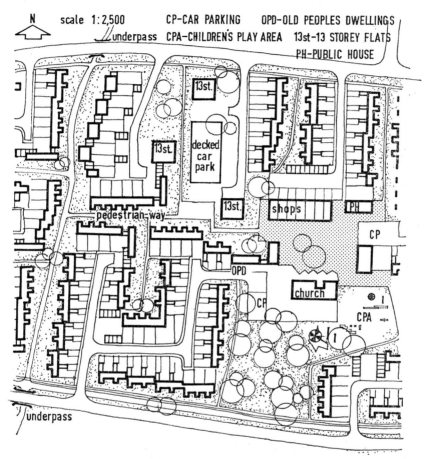

Figure 19. Part of the Almond Spring Housing development at Stevenage New Town. Layout on Radburn lines with large-scale garage provision and separated pedestrian and vehicle ways.

But in some of the new towns, planners and architects tried to look at this whole question of the car in the residential area in a more fundamental way. The pioneering work on 'Radburn' layouts at Wrexham, Northampton and Sheffield attracted their

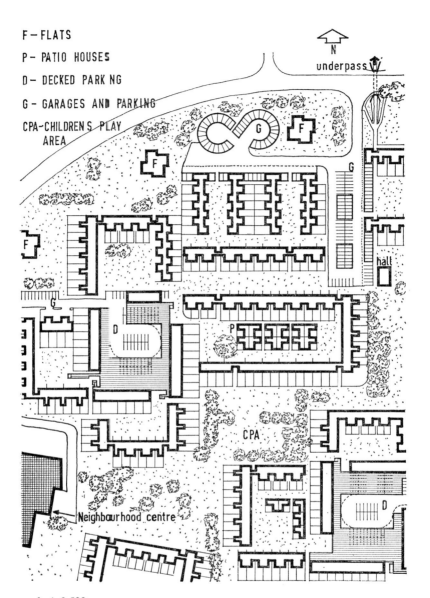

scale 1:2,500

Figure 20. Housing in the Laindon Neighbourhood, Basildon New Town. Garages under pedestrian decks and in double horse-shoe shapes have been used to provide for one car per family but still retain a pleasant, yet urban, development.

104

attention and they began to design their own Radburn-type schemes. One of the first new towns to do this on any appreciable scale was Basildon (Fig. 20) where many schemes were prepared and built using dwellings in 'superblocks' with garages at the ends of gardens and entry into the dwellings from the garage side as well as from the footpath side. Similar developments also took place in other new towns, notably Stevenage (Fig. 19) and Harlow. It is perhaps worth noting that this new thinking about housing layouts did not result in many built schemes before 1965; the more comprehensive layouts on Radburn lines did not appear on drawing boards much before 1960 so that it is only comparatively recently that it has become possible to assess their successes and weaknesses. Nor should it be thought that the new towns were alone in seeking fresh forms of layout on Radburn lines; others—notably in the cities of Sheffield and Coventry— were working along similar lines particularly where flats and houses formed mixed development at higher densities.[1]

The main weaknesses revealed in these schemes, both in the new towns and in city schemes at Sheffield and Coventry, confirmed the difficulties revealed earlier at Wrexham and Northampton. Despite considerable care and attention to detail, it proved very difficult to avoid creating ugly and untidy 'backs'—the service culs-de-sac—which formed a striking contrast with the much neater and often delightful 'fronts' (see Plates 27 and 28). It also became apparent that such radical changes in housing layout could not be successful without equally radical thought being given to the design and internal planning of the houses themselves. In the earlier attempts, insufficient thought was given to the different internal planning which ought to accompany this very different pattern of circulation outside the house. The entry from the car to the house was often along the garden path and through the kitchen; house plans with narrow frontages, which had been developed for reasons of economy on the advice of the Government in the 1950's, were thus being used in greatly different circumstances. The gardens were also narrow and almost invariably separated by wire fencing. The placing of

[1] See Chapter 6.

105

garages along service culs-de-sac between houses tended to improve outdoor privacy by separating and screening the back gardens of houses facing each other, but the narrowness of the

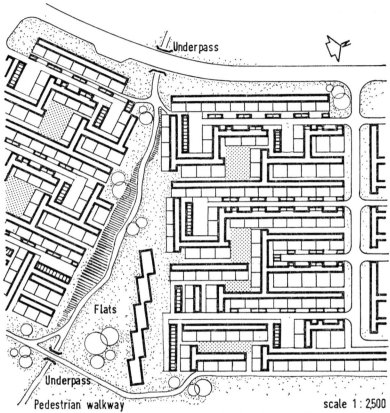

Figure 21. Wide-frontage houses at Carbrain, Cumbernauld New Town. One garage is provided for each dwelling.

gardens made them difficult to arrange; little more than a path to the house and a place to hang out washing remained. To overcome these weaknesses, long-frontage houses were developed, especially at Cumbernauld, providing squarer-shaped private gardens screened at the sides as well as the back (Fig. 21). At the same time the long-frontage house made it possible to have house plans which provided for entry into the house from 'front' or 'back' without going through one of the rooms.

Study of these early layouts has now revealed that, while it is possible to achieve substantial segregation between pedestrians and cars, a more satisfactory environment may result if the relationship of the garden to the house is reversed. If the small

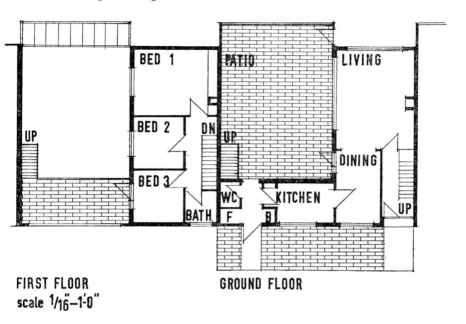

FIRST FLOOR
scale 1/16"–1'0"

GROUND FLOOR

Figure 22. Patio houses, showing L-shaped plans linked in terraces, suggested by Walter Segal in his book *Home and Environment* published in 1947.

private gardens are on the footpath side, instead of facing garage and service courts, they can be combined with the landscaped open spaces and pedestrian ways, away from the service side of the house which once more becomes a street which can be likened to a modern version of the ever-popular mews courtyards in Chelsea and Bayswater. Such courts can be unashamedly devoted to the vehicle, its garaging and parking and to service access. The treatment of the service side of the house is still a formidable

107

architectural problem but not so difficult to solve if the garden is on the other side of the house. Finally, in this consideration of significant changes in house designs, there is the patio house. The L-shaped plan, usually of a single-storey dwelling around a small, enclosed garden or patio has long been beloved by many architects although archaic building regulations have mostly prevented its application in grouped housing in Britain. As long ago as 1947, Walter Segal, in his prophetic book *Home and Environment* gave a number of interesting plans of patio houses (Fig. 22); but few such houses were built in Britain although the advantages which they offer in privacy would seem to be obvious. A major break-through for the patio house came in the winning scheme initiated by the Harlow New Town Corporation in 1961 (Figs. 23, 24 and 25). Here, on a sloping site, there is a carpet of L-shaped patio houses mostly looking out over landscaped greens. There are hardly any roads on this thirteen acre site and garaging in the proportion of one garage or parking space per dwelling is provided under a central podium giving complete separation of pedestrians and vehicles.

Often, with patio houses, as well as some two-storey terrace houses, there are advantages in grouping garages or parking spaces and siting them near the houses rather than building them into the house. This is another lesson from experience. Provided that walking distances are kept short it is possible to group garages or car ports and substantially reduce the amount of road needed. Such arrangements also make it possible to give all houses in a scheme the most favourable aspect for sunlight. If attention is paid to screening gardens, and positioning ground-floor windows on the footpath side of the house, a high degree of privacy can be obtained even though, for two-storey housing, the density is comparatively high–sixty to seventy persons to the acre. This is shown particularly by such schemes as that at Seafar, one of the housing areas at Cumbernauld New Town.

The examples mentioned have been schemes in which a public authority has taken the initiative. One scheme which pointed the way forward for private enterprise was the winning design by

12. Houses at Cumbernauld. The separation of pedestrians and vehicles gives safe areas for toddlers to play, increased privacy and less noise from vehicles.

13. One of the first schemes in the New Towns which began to separate pedestrians and vehicles. Houses and corner-flats along a pedestrian way at Basildon New Town. The scheme was completed in 1960.

14. A pedestrian way at Cumbernauld flanked by existing trees, retained in the rew scheme. The footpath leads to an underpass for the safe crossing of a busy main road.

15. An underpass at Cumbernauld linking housing areas to the town centre.

Clifford Culpin for a housing competition in 1961, conducted by the Royal Institute of British Architects and the magazine *Ideal*

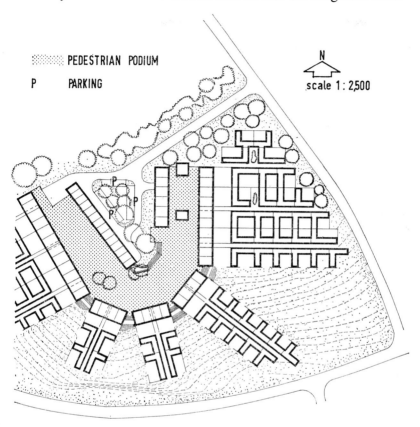

Figure 23. Patio houses grouped round a pedestrian podium with garages under. Prize-winning scheme for Harlow Development Corporation by Michael Neylan.

Home on a site at Harlow New Town (Fig. 26). Here, the architects produced a scheme which, for private enterprise housing, was of a relatively high density–sixty persons per acre–and with a very high proportion of garage and parking provision–137 parking spaces and garages for seventy dwellings. The aim was to

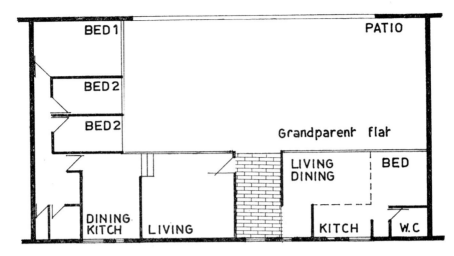

Pedestrian   Lane

scale 1/16″–1′·0″

Figure 24. Plan of patio house with grandparent's flat incorporated
(see layout on page 109)

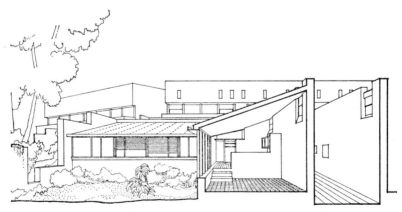

Figure 25. Drawing of patio house in Fig. 24 showing view outwards
to open space (see layout on page 109).

110

avoid the rear access yards of the English variation of the Rad-burn scheme. Instead, the fronts of the dwellings have been planned to be urban in character and to provide for all services to the houses and the whole process of delivery and collection.

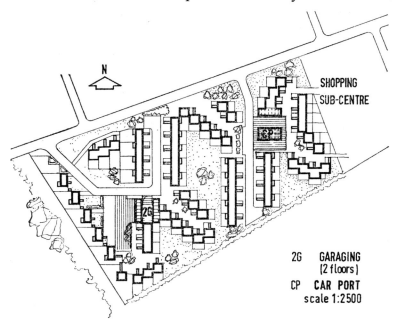

N

SHOPPING
SUB-CENTRE

2G   GARAGING
       (2 floors)
CP   CAR PORT
       scale 1:2500

Figure 26. Private enterprise housing at Harlow: the winning design in the 1961 competition (Plate 16).

In other words, the front of the house is the hub of all activities while each family has a private garden court at the rear which looks out on to a community open space. There is also, within the scheme, a community car port as a further experiment in housing the motor vehicle.

NEW TOWNS ELSEWHERE

Our discussion about new towns has been concentrated on British experience because this forms a case study of the application of theoretical concepts and gives the opportunity to learn from these experiences. But the British new towns are not the only examples of new urban communities being developed. All over the world, especially in the last twenty years, there have been new, planned settlements, varying in scale from the new capital city of Brasilia–described by Victor Gruen as '. . . the most gorgeous autocratically planned new city in the Western world, the construction of which has nearly succeeded in bankrupting a nation'[1]–to the Canadian company town of Kitimat, a private venture by the Aluminium Company of Canada. A general description of all these ventures and an assessment of their relative significance does not fall within the scope of this volume. Instead, examples have been selected–principally from Sweden, the United States and India–which serve to illustrate the approach to new town building in different parts of the world.

Since 1945, the broad plan for the development of Greater Stockholm has included a number of new satellite towns within the city limits. These new towns, like Vallingby on the western side of the city and Farsta in the south, are linked to the central city by a rapid-transit underground railway system.

The British new towns were already under way when Vallingby was being planned and there is little doubt that the Swedes learnt much from British experience. But Vallingby, which was virtually completed in four years, also had much to teach; its influence on town planners, architects and road engineers in Britain has been profound. Vallingby itself has a population of 24,000 but also serves as a centre for shopping, offices and entertainment for several smaller satellite suburbs of Stockholm; eventually these suburbs, together with Vallingby, will have a population of about 100,000 people. The centre of this 'new town' in the suburbs of Stockholm is a multi-level shopping and commercial precinct centred on a tube railway station which connects to the middle of Stockholm, twenty-five minutes journey away (Plate 18). Escalators bring passengers to the shopping levels, and car parking for 2,500 cars has been provided. The whole layout of the residential areas of Vallingby and their connections with the 'town centre' shows a great advance on

[1] Victor Gruen, *The Heart of our Cities*, Thames and Hudson, 1965.

16. Houses for sale at Harlow New Town. The winning entry in an important competition in 1961. The large, open car port and the individual garages adjoin the road, leaving pedestrian ways, patio gardens and open space free from traffic (see layout on page 111).

17. A cluster of larger houses for higher income groups at Basildon New Town. The houses, completed in 1962, were designed as a group so that they relate to each other but with variety to suit differing requirements (see layout on page 102).

18. Vallingby, a satellite town in the suburbs of Stockholm. A multi-level shopping and commercial precinct straddles the railway. Housing is grouped round the centre in neighbourhoods of between two and three thousand people.

View of "Platform" pedestrian level

19. A satellite town for 25,000 people at Erith, Kent, designed by town planners, architects and engineers of the London County Council. As it is to be built on land liable to flooding from the near-by Thames, all housing is built off pedestrian platforms twelve feet above the ground. Cars for every family are to be garaged under the platforms.

British towns. Within the housing area are 'neighbourhoods' of about 2,000 to 3,000 people each, in turn subdivided into groupings of houses and flats around road spurs and culs-de-sac. Within the housing 'superblocks' are the footpaths with underpasses or overpasses where they cross motor roads. Garages were provided for two families out of every three. An underpass for pedestrians connects the nearby main residential area with the town centre.

In earlier chapters the special contribution of the United States of America to planning thought was emphasized. The Radburn idea has been a key contribution to modern community planning in the motor age. Yet, unfortunately, there is not yet one completed new town in the U.S.A. that is sufficiently comprehensive to be worthy of the name. The nearest so far have been the Greenbelt communities built between the wars. These, as one American critic has commented: '. . . are still more pleasant places to live than most suburban areas that were to come. But they are part of suburbia and not, strictly speaking, new towns. They never succeeded in attracting the intended industries.'[1]

More recently, big industrial and commercial concerns in the United States have sought to fill the gap in Federal or State enterprise by promoting new towns. One such proposal is for a new town of Columbia, with a proposed population of 110,000. This new town is backed by the Connecticut General Life Assurance Company. The most developed of these schemes for new towns that are to be financed by private concerns is for Reston, a proposed town for 75,000 people on a site in Virginia, eighteen miles from Washington. It is proposed that Reston should be a balanced community with not only residential areas and shopping but also industry. The plan places strong emphasis on the pedestrian and on linking together a series of Radburn-like clusters of dwellings. Instead of the more familiar American suburban pattern—of a grid of separate, detached houses—it proposes residential clusters of dwellings with variety in the

[1] Wolf Von Eckardt, 'U.S.A.: The Case for Building 350 New Towns', *Town and Country Planning*, January, 1966.

dwelling types, including groups of terrace houses as well as an occasional tall flat block for emphasis. Generally, the residential density will be relatively low but there are to be 'sinews' going up to a higher residential density of sixty persons per acre.

The main feature of the plan is the concern for recreation. Not only is the town to be divided into seven 'villages', each with its own shopping centre and its own character, but each village is also to have its own recreational facilities which are conceived on a lavish scale, with residential areas designed around particular leisure hobbies, such as boating, horse riding and golf. One is left wondering whether this is really likely to be a new town for the ordinary American citizen or pleasure-orientated houses for very affluent American upper-middle-class families. With five golf courses, several very large yachting and boating lakes, marine clubs, riding schools and stables, it is obviously the intention to cater for increased leisure on a grand scale. Despite the reservation for industry of a substantial area of land adjoining the existing freeway which cuts across the middle of the new town site, one wonders whether factory workers are likely to be included in the 'balance' of this community and enjoy freely the fruits of this residential playground.

Reference has also been made in earlier chapters to early Russian ideas and plans. More recently, town planning has been receiving greater attention in the countries of Central and Eastern Europe as well as in Russia. Following the initial but long period of reconstruction after the Second World War in Europe, many of these countries have begun to turn their attention to large new settlements including new towns. In Czechoslovakia, for example, a new mining town for 40,000 to 50,000 people was begun in 1952 at Havirov in the Ostrava District.

Town planning in Russia has also turned to the development of satellite new towns on a considerable scale. The neighbourhood concept is clearly identifiable in Russian town planning with the 'micro-district' as a residential unit built around local social, cultural, educational and shopping facilities. The first of the satellite towns for Moscow, for example, was begun in the late 1950's. This is the town of Sputnik, planned for a population of 65,000 to 80,000 people. Its residential micro-districts are to have about 6,500 population, each with district centres serving several micro-districts.

Unfortunately, too little is known in Western Europe about the realization of these parallel efforts in the countries of Eastern Europe. If this experience is to be more widely valuable then the weaknesses must be frankly discussed and the lessons learnt from these as well as from successes. Communication links between town planners and architects of Western and Eastern Europe are still, unfortunately, beset with overtones from the international political arena as well as the difficulties of comparison due to differences of technical standards and methods of measurement; so that objective evaluation of this experience by British planners becomes all but impossible at present. Nevertheless, at some stage, it must be attempted but with a need to make assessments on a relatively sophisticated basis making due allowances for differing economic, social and cultural contexts.

These provisos are double necessary when considering the new town of Chandigarh in the Punjab, India, because of the relative difficulties of a developing country. In 1947, partition between India and Pakistan left the East Punjab without a state capital. In 1951, an international team began to prepare a master plan. The team was led by Le Corbusier and included Pierre Jeannerat, Maxwell Fry, Jane Drew and Indian town planners and architects. The virgin site, which had already been chosen, was bounded on the north by the foothills of the Himalayas and by monsoon streams on both east and west; it was only to the south that there was any scope for further expansion if the original target population of 150,000 was going to be exceeded later.

The plan takes the form of a grid of rectangles, roughly half a mile by three quarters of a mile each. Most of these rectangles are the sectors which are designed as residential areas–neighbourhoods with the provision of everyday needs and amenities within walking distance–but they are linked together by an open space system which flows through them in a north-south direction. This open-space system provides connections with the town

centre and the group of government and administrative buildings
to the north of the town (see Fig. 27).

Circulation within Chandigarh is provided by a hierarchy of
routes, classified according to purpose, from the main town
distributor roads down to the cycle tracks through the green
spaces. Access to the town distributor roads from the residential

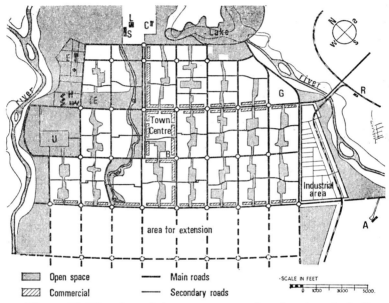

Figure 27. The plan of the new town and regional capital of
Chandigarh in the Punjab, India.

areas is limited so that there is minimum interruption of traffic-
flow by minor intersections and as much pedestrian safety as
possible. It would, of course, be possible to criticize the approach
by applying European standards of car ownership and then to
criticize the plan because the junctions in the grid road system
are all one-level roundabouts and minor routes and pedestrian
ways cross the roads at the same level. Yet such criticism would
reveal naïvety about the economic and political position which
circumscribed the designers. The purist may insist on nothing less
than 100 per cent separation of pedestrians and vehicles regardless

of the fact that no European new town or existing city has so far achieved these ideal circumstances in a deliberate planned way. But the creation of the pedestrian-dominated 'sectors' at Chandigarh is sufficiently significant and advanced for a new city in an emerging country.

A simple system of planning control operates at Chandigarh to integrate buildings with each other within the commercial and residential areas. All but the simplest buildings must be designed by architects. These designs are considered in all aspects by a team of chief local government officers who meet weekly for this purpose. All dimensions are related to a modular system introduced by Corbusier. This results in a series of standard sizes of house plots and shop frontages. Dimensions from the back of the pavement to the façades of houses are also defined together with the total area of the site to be covered by buildings and the prescription of floor and roof levels. This system may appear to British minds to be somewhat bureaucratic and inflexible; in fact, its operation is speedy, realistic and humane. Its success is proved by the rate at which Chandigarh has grown to nearly 150,000 population already whilst avoiding the architectural anarchy which would undoubtedly have arisen if these standards and controls had not been imposed. The result is not always a high standard of architecture in the individual buildings but the order exists from which truly great civic architecture may develop.

Chandigarh has to be visited to savour its real colourfulness, to view the achievements of the new city against the background of its bustling life. So much has been achieved in a short time, not only the individual buildings—the powerful Corbusian Legislative Assembly, Secretariat, High Courts, the University—but the schools and housing by other architects as well as the great pattern of roads, sanitation and water supply and the already mature landscaping.

EXPANDING EXISTING TOWNS

Since 1952 (1957 in Scotland) there has been legislation in being in Britain to help in the rapid expansion of existing towns.

This provides for financial assistance from government funds to local authorities that agree to take population from major cities that are seeking a solution to overcrowding and congestion. So far the operation of the Town Development Act and its Scottish counterpart has resulted primarily in agreements between London and receiving authorities in South Eastern England and Glasgow and receiving authorities in Scotland.

In the case of Glasgow, the original Clyde Valley Plan, prepared by Abercrombie and Matthew and published in 1946, drew attention to the overcrowding in Central Clydeside and particularly in Glasgow itself. At that time, there were 700,000 people living in 1,800 acres of central Glasgow with residential densities of more than 400 persons per acre. In the Clyde Valley Plan it was suggested that half the city's population should be moved to new areas beyond Glasgow's city boundaries. Since then the new towns of East Kilbride, Cumbernauld and Livingston have been providing a partial solution to these problems. But there was still a great deal of scope for, at one and the same time, reviving many small Scottish towns that had declining fortunes while also helping to solve Glasgow's acute housing problems. Now, more than fifty local authorities in Scotland have made agreements with the Glasgow City Corporation to receive families, including larger towns like Perth and Hamilton, as well as many smaller towns. In the main, these schemes are adding several hundred and up to about a 1,000 dwellings in the receiving towns. But even this scale of expansion will mean, in many cases, very substantial increases in the size of the towns concerned and could mean a radical alteration in their prosperity if industrial firms can also be persuaded to go there. It seems as though these schemes are likely to make the most rapid progress in the areas of Central Scotland, between Glasgow and Edinburgh, where government plans for regional development have identified particular areas with the greatest potential for growth and where industrial development is therefore especially likely to be encouraged.

The problem of dispersing London's population over the wider area of Southern and South East England is a much larger and a more complex problem. It is not so much a question of reviving declining towns as of making planning decisions as to which towns should grow faster than others–that is, selecting towns for expansion for sound planning reasons connected with their potential and in a well-conceived regional and sub-regional context. Many of the town expansion schemes in Southern England so far–almost all of them fostered before any deep consideration at government level was given to regional economic or physical planning contexts–seem to have such a casual planning basis that any likelihood of their being the best planning decisions could be regarded as coincidental. Thus, Mr. Jack Whittle, the architect and town planner largely responsible for town expansion in the former London County Council has said, of the experience of the new towns in Southern England: 'Almost without exception the twenty-one town expansion agreements in which the London County Council are partners have started as a result of an approach by the receiving towns themselves. Expansion has not been thrust upon any town; they have themselves sought it, and those authorities with the most enthusiastic members and officers by and large proved the most successful. Men who build towns, the attributes of the town planners and builders, and the men and women on committees, are as potent a force in creating conditions for growth as any more concrete planning or economic considerations. It almost seems that given energy, enthusiasm and goodwill, town expansion schemes can be a success any-where.'[1]

It was local enthusiasm and local pride which built the industrial towns of Britain in the nineteenth century in a situation where expansion was in the air. But it also created the bad environment, bad housing conditions and bad working conditions of nineteenth-century industrial Britain. The lesson is surely not that we should continue with only local enthusiasms and parochial expansionism as our guides but that the location of expansion must now be put into a wider context. Thus town expansion schemes should relate to regional and indeed national

[1] Jack Whittle, 'Folk–Work–Place; Dispersal in Practice', *Official Architecture and Planning*, December, 1964.

priorities and should become part of a more efficient sub-regional framework. It is not only the relationship between the expanded town and the exporting metropolitan authority which needs to be considered but the relationship of the expanded town to its whole sub-regional area. This might lead to proposals for a whole group of linked town expansions, perhaps with a large new town to provide a major centre and with full consideration being given to the services required for the group of towns as a whole and to the road and rail links between the towns and with the national networks.

This wider approach has been started with such studies as that of Northampton, Bedford and North Bucks.[1] Instead of merely considering the expansion of Northampton in isolation, the brief of the planning consultants was to consider the possibility of the large-scale expansion of a group of existing towns. It is now proposed that a whole new 'city-region' be developed, based on the expansion of Northampton and Bedford but also including the large new town of Milton Keynes. Northampton, with a population of 120,000 at present, is to expand to 222,000 by 1981 and 300,000 by the year 2000 while the new city of Milton Keynes is planned to reach 70,000 population by 1981 and 150,000 by the end of the century. But the proposals for a Great Northampton city-region also include substantial expansions of Bedford, Wellingborough, St. Neots and Banbury. The aim is a new city-region, a group of towns and cities roughly midway between London and Birmingham, on the national lines of road and rail communication, strong enough to act to some extent as a counter-magnet to the two major conurbations.

Whenever the attempt is made to put town expansion in Southern Britain into context one returns to the problem of determining an overall strategy for the planning of South-East England and of London. *The South East Study*[2] began the process of looking at this problem in a wider context. From this study emerges proposals for a series of substantial, Government-

[1] *The Expansion of Northampton*, H.M.S.O., 1966.
[2] *The South East Study*, Ministry of Housing and Local Government, H.M.S.O., 1964.

planned, urban developments at some fifty to ninety miles from the centre of London. Around such towns as Ipswich, Peterborough, Northampton, Southampton-Portsmouth and Ashford, it is proposed to develop new clusters of urban development

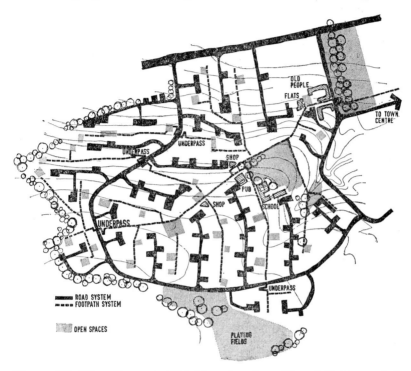

Figure 28. The Clements Neighbourhood at Haverhill Expanded Town. Complete separation of vehicle and pedestrian is made possible by the use of five underpasses.

sufficient to absorb an additional one million people during their first twenty years of development.

These proposals go much further than the earlier large town expansions fostered by the old London County Council through the Town Development Act procedure, at centres like Basingstoke and Andover. Nevertheless the achievements of town expansion on the scale of Basingstoke, for example, should not be forgotten. Under the agreement signed in 1961, between the London County Council, the Hampshire County Council and the Basingstoke Borough Council, the town began its process of trebling in population–from 26,000 to 76,000 by 1976–a period of fifteen years. The number of dwellings that will eventually be built at Basingstoke–11,500–is equivalent to the whole of the town expansion agreements made in Scotland by the city of Glasgow with some fifty-six local authorities up to the end of 1966.

NEW TOWNS: SECOND SERIES

The second series of new towns in Britain began in 1956 with the designation of Cumbernauld, the fifteenth of the British new towns started since the end of the Second World War.

Planners of Cumbernauld were given the honour–and the responsibility–of drawing together the lessons of earlier new towns and attempting to go one stage further. In one very important aspect they succeeded. They re-thought the whole question of the relationship of the motor vehicle and new towns and produced a complete town layout with pedestrians and vehicles largely separated. With Cumbernauld, pedestrian and vehicle separation took a gigantic leap forward in British planning. At the same time the new town centre was conceived in quite a different way to that of earlier new towns, as a great complex of buildings and different uses at different levels, with the servicing of buildings and car parking beneath the main shopping and pedestrian 'decks'–the great ocean liner was the model of the elongated town centre of Cumbernauld and for many of the other new towns that are to follow.

In another respect also Cumbernauld broke new ground. There was a recognition of the importance of the effect that the location of industry had on traffic at peak hours. The siting of the main industrial areas was split between two opposite ends of the town. Moreover, some small and light industry, particularly firms employing part-time female workers, was located close to residential areas.

The new towns that followed Cumbernauld are carrying further many of the principles of new town design and adding others as well. The separation of pedestrians and vehicles is now widely accepted and applied. There has been a tendency for residential densities to fall to less than those tried, for example, for some of the two-storey housing at Cumbernauld but to remain higher than in earlier new towns. This acceptance of slight reductions to lower densities than those technically achievable is partly due to the acceptance of higher standards of space around dwellings but also to the need to allow additional space for garages, visitors' car parking and toddlers' play spaces.

There seems to have been considerable reluctance, even in the second wave of these towns, to get to grips with the complexities of public transport. Looking through some of the master plan reports, with their much greater awareness of the vital importance of the motor vehicle as compared with the earlier new towns, one almost feels that some of those engaged in new town planning have tried to wish away the need for public transport. There is almost an unspoken assumption that everyone will be able to use a private car at all times.

It was left to Runcorn, whose master plan was designed by Professor Arthur Ling and his colleagues, to evolve a new approach by linking the structure of the town to a public transport system on a separate track. The approach adopted at Runcorn is based on the recognition that even when maximum car ownership is reached, at the beginning of the next century, there will still be a need for some form of public transport (for some of the journeys to work and for high proportions of the journeys for shopping, school and social journeys) either from preference – because it is quicker and cheaper – or because a car is not available to all members of the family for the whole of the time.

The aim of the Runcorn Master Plan is to provide a community transport system which is an integral part of the town's structure and not provided as an afterthought (Fig. 29). The Report[1] discusses the principle of a separate track system and points out that the tram might have survived in Britain if there

[1] *Runcorn New Town – Master Plan*, Runcorn Development Corporation, 1967.

123

had been networks of separate tracks; instead, trams were usually routed along existing roads. When the numbers of cars increased, the inflexibility of the trams, occupying the centres of the main roads, led to further congestion and decreased the capacity of the roads to tackle the increasing number of vehicles. For a time,

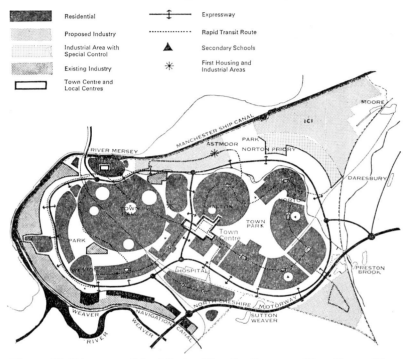

	Residential		Expressway
	Proposed Industry		Rapid Transit Route
	Industrial Area with Special Control	▲	Secondary Schools
	Existing Industry	✳	First Housing and Industrial Areas
	Town Centre and Local Centres		

Figure 29. Diagram of the Master Plan for Runcorn New Town. The rapid transit route forms a separate track—a 'figure of eight' with the Town Centre at the crossing point.

trolley buses replaced trams in many cities and towns but these in turn were superseded by motor buses which could show economies in running costs as well as greater flexibility. But still further increases in road traffic operated against bus transport because the same roads were being used as for all other vehicles. As a result, the speed of bus journeys has been reducing and is appreciably lower than that of the private car; in the centres of big cities, buses average only about 8–10 m.p.h. This has often led to decreasing use of public bus transport and produced a spiral effect of higher fares, less frequent services, elimination of the least used routes and thus a further reduction in the likelihood of such a public transport service providing an attractive alternative to the private car

The solution adopted at Runcorn is to provide a separate rapid transit track linking the communities, the town centre and the industrial areas. Walking distances are cut to a minimum so as to make door-to-door journey times by public transport favourably competitive with those of the private car. It was estimated at Runcorn that at least 50 per cent of work journeys would need to be the objective for the new town's transport system if a cheap, fast and frequent public service was to be provided.

A number of alternative rapid transit systems were considered, including several monorail systems. It was decided that although a form of monorail system might provide a solution for a town or city with a larger population, the capital cost involved precluded it from being used at Runcorn. The system to be adopted utilizes long, single-deck buses carrying eighty to ninety passengers each, operating on a separate road 'track' at an average speed of about 22 m.p.h., compared with the estimated average speeds of only 12 m.p.h. which could be achieved on multi-purpose roads.

Another feature of the second wave of new towns is the reappearance of the neighbourhood concept as the basis for town structure. However, the proposals for Skelmersdale continue in the same vein as Cumbernauld, denying the validity of any neighbourhood structure. The Preamble to the planning proposals refers to '. . . the concept of the town as a compact urban centre with surrounding recreation areas. A large proportion of the population can be within easy walking distances of the central area and all of them can have access on foot to the areas of open space. The development of these principles leads to the abandonment of the neighbourhood system and the central area can thereby be strengthened.'[1] This repeats the approach adopted

[1] *Skelmersdale New Town Planning Proposals*: Report by Skelmersdale Development Corporation, December, 1964.

at Cumbernauld, although this is hardly surprising since the main town planner concerned was L. Hugh Wilson in each case.

In contrast to this, the Runcorn plan is based on a town structure formed around residential groups (Fig. 30). The Runcorn Report states: 'The structure of the town is based on a grouping of communities, each with a local centre. Because of the extent of the town, local centres are essential to cater for the everyday social and shopping needs of the population within a reasonable walking distance.' The Report goes on to claim that: '. . . studies of the relationship of social and commerical facilities to population emphasized educational requirements, particularly primary schools, as basic determinants. . . . A residential community of 8,000 persons gives the most satisfactory balance of local requirements, leaving the larger-scale facilities such as secondary schools and health clinics to be related to two such communities. . . . These theoretical calculations have to be adjusted for varying site conditions and layouts. The community sizes would vary for the same reasons although the margin of variation is restrained by the necessity of ensuring, as far as possible, that the communities are composed of multiples of 2,000 population so as to provide an economical basis for the provision of primary schools.' Thus the structure of Runcorn is to be related to communities of 8,000 which group in four neighbourhoods of 2,000 people each round a local centre (see Fig. 30). This bears a striking resemblance in structure to the original concept of neighbourhood clusters adopted as the structure of Harlow (Fig. 15), although experience has shown the need for each cluster to be on a smaller scale than at Harlow and has enabled the calculation of the relationship between population and social facilities to assume more sophisticated forms.

A similar approach is to be the basis of the structure of Washington New Town on Tyneside–a few miles west of Sunderland and just south of Gateshead. The new town is to be made up of nineteen 'villages', the approximate size of each village–4,500 people or about 1,400 households–has been related by the population required for a two-form entry primary school. Each village is to have a village centre to provide a social nucleus.

126

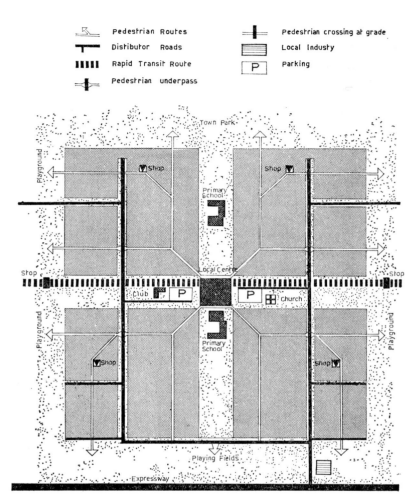

Pedestrian Routes | Pedestrian crossing at grade
Distributor Roads | Local Industry
Rapid Transit Route | Parking
Pedestrian underpass

Figure 30. Diagram of the community structure at Runcorn showing the communities of eight thousand people grouped into four neighbourhoods of about two thousand people each around primary schools and sharing a local centre.

127

The population to be served by each centre will all live within about a quarter of a mile radius from this local centre and will therefore be able to walk to it. The centres are to be sited along the inter-village footpath system. The primary school with its playing fields (for use by children out of school hours and in the holidays) will be close to the centre and connect directly to the pedestrian walkway system. It is not intended that these 'villages' will be rigid, self-contained units. But, as the Report states: 'They are physically identifiable and will help to provide a frame of reference for the many overlapping activities throughout the town.'[1]

The Washington New Town plan is also based, in addition to an acknowledged neighbourhood structure, on the recognition of the need for smaller social units and groupings as well. The Report says: 'Even in an area as small as a village the physical form of development is likely to be divided by the topography, by the pattern of development roads and by the inter-village walkway system into a number of subdivisions. These will vary in size with a lower limit of about 200 dwellings.' This unit is called a 'place' and it is suggested that the place should have a degree of identity through its architectural form and through variations in the services it might include – such as corner shops, telephone booths, post boxes and play areas. Within each 'place' there would also be further and smaller social units – 'groups'. These would vary in size from between twenty-five to fifty families forming physically identifiable groups of dwellings, for example, around enclosed, semi-private gardens, which would be for the exclusive use of the households surrounding them. It is emphasized that these are not an attempt to artificially impose social groups: 'People living in houses round these communal gardens may be expected to have as many friends across the streets as across the garden.' The importance of privacy is also fully recognized: 'Every house should have a private garden, and direct access to the street.'

Thus what is emerging is a counter-trend to the centralized new

[1] *Washington New Town Master Plan and Report*, Washington Development Corporation, 1966.

20. Early Local Centre at The Stow, Harlow, with shops fronting a street but with wide pavements.

21. Later Local Centre at Bush Fair, Harlow. A pedestrian shopping precinct takes the place of the older type of shopping parade on two sides of a busy road.

22. View of a residential area of Skelmersdale, showing patio houses, pedestrian way and a group of small local shops.

23. A playspace in the first of the housing areas at Skelmersdale New Town.

town concept of Cumbernauld. It is not so much 'back to the neighbourhood' and certainly not back to the rigidity of the neighbourhood unit theory. Neighbourhoods become a framework for a new town rather than a formula. Size of area and population will vary with density; and topography will also play a considerable part in determining town form. The concept of overlapping activities replaces that of self-containedness together with the recognition of the importance, from the point of view of social groupings, of smaller units–the housing area of up to 200 dwellings and the smaller group of twenty-five to fifty families which assume greater importance in terms of 'neighbourliness'. The primary schools are, once more, regarded as the principal focus for residential areas but with smaller neighbourhoods of about 2,000 (based on a small one-form entry primary school) or 4,000 to 4,500 people (based on a double-sized, larger primary school system with two parallel sets of children–a two-form entry school).

This approach is well summed up in the Report on the plan for Washington New Town: 'It has always been the aim in new town design to promote social intercourse within the community at various levels. For example the town centre is the focus of some social activities, available to the whole population. At the same time, the need has been felt for sub-centres providing services and facilities for social intercourse for smaller groups–"neighbourhoods" in first generation new towns. . . . The idea of breaking-up a town into separate residential areas with fixed boundaries, providing a range of services at a certain level for the whole of their population, is no longer socially very relevant. Instead it is recognized that each family or individual will probably have social contacts at a whole range of scales varying from small groups surrounding the home upwards to facilities in the region outside the new town itself. . . . Instead of thinking of the town as composed of a number of elements–industrial estate, housing estate, central area, etc.–each circumscribed by a fixed boundary and connected together by a communications network custom-built for the purpose, the town is now seen as a rather more complex, overlapping structure. The emphasis in design is away from division into separate compartments and towards the creation of a structure for the town which will allow the various functions to adjust to social and economic change and which will encourage diversity of opportunity for its residents.'[1]

FURTHER LESSONS

We have tried to show how the first set of new towns were developed in Britain and the lessons that were learnt from this. It has constantly to be remembered that those who planned these early new towns had very little theoretical study on scientific lines or practical experiment to rely on. The surprising fact is not that mistakes were made but that these mistakes were not more serious ones and that the new towns have turned out as well as they undoubtedly have, within the limits which circumscribed the designers and planners. Weaknesses have been revealed in the crude theoretical concepts applied and results have often proved to be disappointing in some degree. But some of these faults lie not so much in the original concepts as in the way these were turned aside, whittled away, emasculated or countered by other short-term considerations. Certainly financial policy is partly to blame for a host of minor weaknesses which together add up to a substantial part of the shortcomings. Those extra few pounds per head were often not available when needed, resulting in mediocre rather than much better results. All the way along the line the new towns were made far less successful than they might otherwise have been by the 'pinchpenny' attitude with which they were saddled. The late Lord Beveridge, who was then chairman of the New Town Development Corporation at Newton Aycliffe, commented: 'One after another, ambitious schemes for making the perfect town fell by the way or were delayed. At Newton Aycliffe we hoped to make a smokeless town with district heating, but we found the costs so high that we could not make it a compulsory addition to the rent of every person in the town. We wanted to build a good theatre for com-

[1] *Washington New Town Master Plan and Report*, Washington Development Corporation, 1966.

munity drama but the cost was frightening. We thought of a first-class hotel, but we found ourselves reduced to a public house with no rooms for anything but drinking.'

The flourishing of community interests and social life in the new towns have often been triumphs of determination by the new town dwellers themselves in overcoming formidable obstacles. A survey, which attracted considerable attention at the end of 1960, into the 'Use of Leisure' in Crawley New Town, showed how inadequate were the social and community facilities available and the difficulty, in such conditions, of being able to draw lessons about social habits in a new town: 'The present uses of leisure must be seen against a background in which the only premises in the centre of Crawley in which adults can foregather for voluntary activities are those which served before the war for a community one-tenth the size of the present. Compared with the well-designed new house, factories and schools all these places are unattractive and they repel all but the most hardy enthusiasts. At the time of the survey the Central Library was housed in the remnants of a bombed chapel though it has since moved to the relative elegance of a disused Sorting Office. Given these conditions it will remain impossible to discover whether New Town conditions can lead to increased community responsibility or to greater participation in cultural activities.'[1]

The distinguishing feature of the second wave of new towns has been the grappling with the challenge of the motor vehicle, the firm establishment of the principle of separation of pedestrians and vehicles, and the beginnings of attempts to provide a satisfactory balance between private and community transport. But within this second phase of new town building there has still been relatively little advance beyond the first phase in conceiving of new towns as more than a reception area for the 'overspill' of population and industry. Arising out of the study of this second wave of new towns there has undoubtedly been a greater realization of the need to relate the new town to nearby towns and cities and to consider in more detail the place of the new town in

[1] Letter in *The Times* from the Chairman of the Crawley Branch Workers' Educational Association, 23rd September, 1960.

a complex of larger and smaller urban settlements into which it must be fitted. But frequently, there appears to have been a 'cart before the horse' position in that the new town's location, size and area was determined *in advance* of any wider and deeper regional and sub-regional study. A classic example of this unsatisfactory process at work is the case of Dawley New Town in the West Midlands. A hurried study of the practicability of locating a new town on a difficult site at Dawley was made and the designation of the new town proceeded. But after several years of effort it was then decided that it would be better to take in the adjoining urban areas of Wellington and Oakengates, thus requiring the reformulation of the whole basis for a new town now to be of much larger size.

The problems created by this topsy-turvy kind of approach are also illustrated by the example of Washington New Town, which was proposed as a new town arising out of the Hailsham Report on regional development—*The North East: A Programme for Regional Development and Growth*—issued by the Government in November, 1963. It is virtually impossible to separate Washington from Gateshead, only half a mile from the boundary of the new town designated area, or from Sunderland which is only one and a half miles away. The key to the inter-relationship of these urban areas is the journey-to-work movement within a complex of urban units. In these circumstances, no new town could be self-sufficient. An effective mass-transit system of public transport is required. Yet, as the Report for Washington reveals, the plan for the new town had to be prepared without this wider framework. As the Report states: 'In preparing the Master Plan for Washington it has been found difficult, indeed almost impossible, to predict the future pattern of public transport in the urban region. Until very recently there was no body charged with the task of considering this problem for the region as a whole. Now the Northern Economic Planning Council has established the Study Group on Transport Integration for the urban region. This group has only recently begun work and no conclusions have been forthcoming in time to make a contribution to the Washington Plan. In the absence of a regional plan, conservative assumptions have therefore been necessary as to the pattern of public transport in the region. . . . It is to be hoped that the Northern Economic Planning Council will press forward study of regional transport problems as rapidly as possible. Unless firm proposals are forthcoming very soon it will be difficult to take account of them in the development of Washington.'[1]

Slowly however, the lessons are being learnt. Starting with the new town of Livingston in Scotland, begun in 1963, greater efforts are now being made to study regional and sub-regional areas before determining the location, size and function of a new town. However, in the case of Livingston, the decisions to set up a new town, and to survey the whole region surrounding the new town area, were taken virtually at the same time. Professor D. J. Robertson the economist/consultant for the West Lothian Survey and Plan—the area within which Livingston is located—argues that there were advantages as well as disadvantages in this: '. . . the regional planner is by virtue of the development which is going on around him forced to put his decisions more firmly to the test of practicability, since the regional plan may be a basis for action even as it is being formulated. . . . The new town planner will feel the constraints of working with another professional team looking over his shoulder, and in circumstances in which his plans may be subject to re-examination in the light of possible regional proposals; but he gains the prospect of a development of the region around him in harmony with his own proposals.' At the same time Professor Robertson admits that the active development of the new town, which is taking place at the same time as the regional survey: '. . . will create a need for compromise which may not be consonant with the proper development of a plan for the whole region.'[2]

Nevertheless, the achievements in Scotland, in putting new towns and town expansion into a regional and sub-regional context—to a far greater extent than has so far taken place in England

---

[1] *Washington New Town Master Plan and Report*, Washington Development Corporation, 1966.

[2] D. J. Robertson, Editor, *The Lothians Regional Survey and Plan*, H.M.S.O., 1966.

and Wales–should not be under-estimated. The West Lothians Regional Study and the plan for Livingston New Town rest on a foundation built up over many years by the regional work and studies of the Scottish Development Department which has led the way in Britain in developing regional planning as the basis for more detailed physical planning proposals. It is to be hoped that the Feasibility Studies for possible national growth areas in Britain–Humberside, Severnside and Tayside (the Dundee area)–will mark a new phase in the formulation of regional proposals and provide the basis for the planning of new and expanded towns within the context of the regions and sub-regions in which they are to be located.

CHAPTER 6

Renewing Towns and Cities

The value of the new towns and their influence on town planning generally have been emphasized in the previous chapter. New towns, and the expansion of selected existing towns, have proved to be valuable means of meeting the great population expansion with which countries all over the world are faced.

But the contribution which new towns in Britain have made towards solving the housing problems of the great cities and towns should not be exaggerated. By the end of 1966 160,000 houses had been built in the new towns but this was less than 3 per cent of the total number of houses built in Britain since the end of the Second World War. Even if we include the expansion of other towns, the provision of new towns and town expansion is nothing like sufficient to meet the housing shortage of even the minority of cities and towns which have been able to look to these means of relief. Councillors, administrators, planners and architects have therefore found themselves more and more involved in *replanning* schemes, large programmes of rehousing to deal with obsolescence and decay, as well as continuing slum clearance. These measures have meant not only the provision of new dwellings to replace those that have to be demolished but also the readjustment of the pattern of industry and commerce, shopping and entertainment, to meet changing habits and demands. Above all the impact of the motor vehicle on the urban environment has produced much more complex problems requiring solution. It is the purpose of this chapter to examine the steps taken so far along this difficult path.

RENEWING TOWNS AND CITIES

How Many Homes?

The renewal of cities has been dominated by the need for great increases in the number of individual homes. In the nineteenth century, industrial towns and cities in Britain grew like mushrooms. Vast areas of our present cities and towns are still covered by this nineteenth-century housing. The earliest and the worst of these are slums—the bulk of the housing which is more than 100 years old. The efforts to sweep these away only really started in the 1930's; the pace of slum-clearance has never, so far, made appreciable inroads into the rate of decay and obsolescence. There is often a feeling that slums can be swept away once and for all. This static concept is highly misleading. The slum dwelling is a product of obsolescence. Unless renewal is proceeding a great deal faster than obsolescence then the slum problem remains as buildings reach the end of their useful life. And this is what is still happening in most of the older urban areas throughout the world.

What does all this amount to in terms of the number of houses required in Britain? It is only comparatively recently that it has come to be recognized that a large proportion of new housing likely to be built at current rates of building is likely to be absorbed by the number of new households which are being formed more rapidly than earlier in the century. Not only is the population increasing with greater rapidity than was expected—through increased birth rates and decreased death rates—but as young people leave home earlier, get married earlier and have children earlier, so the number of new households requiring separate dwellings increases even more rapidly. There will also be a great many more single people living alone to be catered for. Professor Donnison[1] has estimated that while the population increase for Britain is likely to be in the region of from 51 million in 1960 to nearly 65 million by 1990, the number of households may rise from 16·31 million to as many as 21·94 million—an increase of 5·63 million in the period from 1960 to 1990. Thus to

[1] Professor D. V. Donnison, *The Government of Housing*, Penguin Books, 1967.

136

cater for these additional households over the thirty year period, an average of 190,000 *additional* dwellings per year would be required. The current rate of building houses is now about 340,000 dwellings a year in Britain.

The problem of estimating a realistic total of new dwellings required must also, of course, take into account new dwellings required to replace older ones. Out of the 15 million dwellings that are occupied today in Britain, approximately 7 million were built before 1914. About $2\frac{1}{4}$ million of these are more than 100 years old and another $1\frac{3}{4}$ million are more than seventy-five years old. Included within these older dwellings will be nearly all of those that are classified officially as slums.

Slum clearance began in 1930 and 341,000 houses had been demolished by 1938. Then there was a complete break during the war and its aftermath, until 1954; the building programmes after the war and up to this date were almost entirely devoted to new housing. In 1954, local authorities were given the task of preparing fresh estimates of houses unfit for habitation in their area. Altogether they estimated just over 850,000 slums of which they aimed to clear 378,000 by 1960. But achievements did not match even this limited target; by the end of 1960 local authorities had only dealt with about 260,000 houses. Complacency about this state of affairs was revealed in the Government report on this first five years of slum clearance by local authorities: 'Although this falls short of the target they set themselves, it is nevertheless an impressive achievement. In many areas the job has been done or, at any rate, reduced to small proportions.'[1]

But was 850,000 houses the real measure of the size of even the urgent replacement programme for slum dwellings? As Professor Donnison has said: '... despite repeated efforts made by the Ministry of Housing and Local Government to persuade local authorities to give more accurate estimates of the number of unfit houses in their areas, it is clear that the returns they send in are influenced by all kinds of diverse factors.... Many authorities are reluctant to look beyond the numbers of slums they are administratively and financially capable of dealing with during the next

[1] Report to the Ministry of Housing and Local Government, 1960.

few years; some are more aggressive and ambitious and return very high figures; others are reluctant to give their cities a bad name ... the local authorities' estimates of the total number of houses in their areas which are unfit vary without convincing justification. Although, when taken together, they furnish an approximate guide to the regional distribution of the problem, they provide neither a reliable measure of the long-term task confronting each authority nor a reliable measure of the national problem.'[1]

The real size of the housing problem in Britain, if it is conceived in terms of obsolete housing conditions, is perhaps as much as one-third of the total national housing stock. It would seem reasonable to assume that dwellings should be replaced after a maximum life of sixty years. But to bring the normal life of houses down to eighty years in Britain would require 300,000 houses a year to be demolished up to 1981.[2] By comparison, the clearance of slum dwellings had only reached about 70,000 dwellings a year by 1966.

To this scale of replacement should be added the number of dwellings needed by additional households that will be constantly forming–probably a further 190,000 a year for the whole of Britain. Additional dwellings will also be needed to replace those lost through demolition because the land is required for other uses (open space, schools, etc.) and especially to make way for new roads which frequently must be pushed through areas of existing houses. Altogether, therefore, an annual programme of house building averaging at least 500,000 dwellings a year is required if the 'housing problem' is to be eventually solved. This compares with the average of 340,000 dwellings a year achieved in the five years 1961–65 in Britain.

The need to build more houses is obviously the task which councillors and the public see as the main task; they are often puzzled by the apparent inability of the town planner and the architect to build our towns and homes at faster rates. But many key factors concerning the rate of house building are really not

[1] Professor D. V. Donnison, *The Government of Housing*, Penguin Books, 1967.
[2] See L. Needleman, *The Economics of Housing*, Staples Press, 1965.

within the province of the town planner or the architect to solve. The finances of urban renewal and housing programmes and the relative priority of these measures, as compared with other pressing needs, play a large part in determining the rate of renewal and rehousing. Moreover, the backward nature of the building industry and its slowness to adopt techniques of industrialized building for housing–so that large parts of houses can be produced in factories in a similar way to motor cars–has been another limitation on the rate of house building over which town planners and architects have had little direct control. Nevertheless, the efforts of architects like Sir Donald Gibson, Cleeve Barr and many others working in government and large public authority concerns has been a powerful influence towards the wider recognition of the use of industrialized building techniques to increase house building rates. Generally, however, architects and town planners in their technical capacities at local government level can only act within the financial and legislative framework. Decisions for action to solve this almost insatiable demand for more dwellings, together with the finances required to make these decisions operative, are firmly in the hands of Government–and mainly Central Government at that.

BLIGHT AND RENEWAL

Blight is much more than buildings deteriorating in their structural condition. It is a symptom of planning conditions not readily adaptable to the needs of the twentieth century. Blighted areas are caused not only by deteriorating buildings but by mixtures of unsuitable uses of land, by inadequacies of space standards, by the compression of uses in areas which are not adequate in size for so many different purposes, by the inadequacies of circulation patterns–the roads and footpaths–for present-day multiple use by goods traffic, private motor cars, buses and pedestrians, together with the inadequate protection of dwellings from traffic noise, vibration and fumes.

Nothing less than *renewal* of these areas is needed. Thus urban renewal implies the gradual re-ordering of uses of land and buildings to meet present-day requirements and estimated future needs; it implies a continuous process of planning and re-building of towns and cities, sorting out uses that have become mixed in unsatisfactory ways and recombining them in new ways, the elimination of excessive noises, smells and atmospheric pollution, especially from the residential scene. Urban renewal implies gigantic efforts to change our mode of living to the new needs of the latter half of the twentieth century and to *keep on changing* so that our environment remains suitable for current needs. Our slums are an outcome of the failure of society to keep pace with technological change; even though it may have come to be universally accepted that every family has a right to decent housing conditions, our relatively affluent nation has so far failed in practice to make this 'the century of the common man' as far as his habitat is concerned. And on a world scale the problem still assumes immense proportions.

The deterioration of buildings, which typifies slums, also has social and community effects. Deterioration has all the symptoms of disease; *urban blight* is vividly descriptive of the way such conditions can spread from the slums into hitherto healthy neighbourhoods and living conditions. The process can be seen at work in any of our urban communities in Britain as well, of course, as in many other countries. As older residential areas are allowed to deteriorate, former residents move out or die off; they give place to the unskilled, the immigrants, the unfortunates and the unemployable as well as the poor in their old age. Substantial houses become subdivided, more people crowd into smaller space, sharing toilet, bathroom and kitchen facilities where these exist. Moreover, for almost a century in Britain, long before any sort of town-planning policies operated, there has been a hotch-potch of ill-assorted and ill-considered non-residential uses taking root within deteriorating residential areas. It is not only the deterioration of buildings, the decay of structures which causes blight to spread; it is as much this mixed usage–noisy factories and workshops, smelly industrial processes, junk-yards and car scrap heaps all mixed in with housing–which is the fertile basis of the further spread of urban blight.

The 'twilight areas' are growing; the inner residential areas, once solid, neat and clean, though dull and visually unattractive, are gradually becoming overcome by blight. They are becoming 'slums' in the sense of dwellings that are inadequate as up-to-date homes for families. The official 'slums' are mere skimmings off the surface of this much greater problem. The real problem of blight is of a magnitude many times greater.

There are, of course, large areas in all towns and cities where houses could be provided with a prolonged lease of life by installing bathrooms and internal w.c.'s, new plumbing and hot water systems. But such internal 'face lifting' must also be accompanied by improvement of the residential environment outside the dwellings. This will mean the removal of the 'bad mixers' among the other uses—the noisy, smelly or untidy industrial or commercial premises. It will also mean the sorting out of road and footpath systems so that pedestrians can walk about safely and parking and garaging be provided without continuing to litter our streets with idle vehicles. Much can also be done in older districts to create sitting-out spaces and toddlers' play areas by eliminating some of the excessive areas of road that abound in such districts; careful urban landscaping, with the planting of trees and shrubs, can also give warmth and vitality to the dull bleakness of many of these 'twilight' housing areas.

This type of detailed and careful treatment of these complex problems is now receiving close study[1] and a number of schemes for revitalization are under way. An example of this approach is the scheme at Newcastle for the revitalization of Rye Hill. Here, a large area of old but soundly-built terrace housing has been chosen for the first of such schemes by the City Council. Many of the houses are large and have been divided into flats. Most dwellings lack the full range of facilities such as private bathrooms and toilets. The district has little in the way of open space and very few garages or parking spaces except for the streets themselves. As has been done in other cities such as Leeds, it is proposed to modernize the houses and flats internally. But, by

[1] *The Deeplish Study*, Ministry of Housing and Local Government, H.M.S.O., 1966.

Figure 31. Two views of a rehabilitation scheme at Rye Hill, New-castle, showing the existing conditions of part of the areas and the proposed new environment to be created.

itself, this has proved to be insufficient to prevent the deterioration of a neighbourhood. The Rye Hill scheme proposes to improve the whole area. There is to be a pedestrian route running through the middle of the area, formed by closing some streets to traffic; this pedestrian route will link the housing areas with a new primary school and around a small, local shopping centre. Garages and parking space will be provided, partly by demolishing some of the houses, together with new children's play spaces and sitting-out areas nearby. Those backyard industries which detract from the pleasantness of the area will be re-located more appropriately elsewhere and space at the back of buildings will be improved by clearing old sheds and buildings (Fig. 31). In ways such as these, many residential districts could acquire a better environment as well as better internal housing conditions to last for a considerable period. But such measures can only be regarded as solutions in the relatively short-term—improvements which are necessary because of the backlog of older housing in Britain which prevents any quicker action in many of the 'twilight' areas for a considerable number of years. The longer-term problems of renewal of the urban fabric of our towns have also to be tackled.

THE DENSITY DEBATE

Reference has been made earlier to new housing standards and house design in the new towns at low densities. But often in redeveloping towns and cities it may be higher densities which are needed. For many years large areas in existing towns and cities have consisted of two-storey dwellings crammed together at densities of thirty or more dwellings to the acre; when subdivision of houses and overcrowding are taken into account, there are often many more than 100 persons to the acre living there at present. Glasgow is an extreme example of high density in Britain because of the prevalence of four- and five-storey walk-up tenements. It has been estimated by the City Corporation that over half of its one million inhabitants lived at densities of over 450 persons to the acre even in the early 1960's. Even if one assumes the very substantial movement of population from the cities–to new towns and expansion of smaller towns–there is still the need, certainly for a long time to come, to retain relatively high densities in towns.

Higher densities do not necessarily mean that all should live in high flats. Cumbernauld and some of the later schemes in other new towns, including schemes for private housing, have demonstrated that it is perfectly possible to have good housing layout and provide 100 per cent garaging at densities of sixty persons to the acre with two-storey houses. If this approach is taken further in towns and cities, by adding in some flats in very tall blocks and four-storey walk-up maisonettes, it is possible to reach comparatively high densities without all families, especially families with young children, having to live in high flats. As a Ministry of Housing publication states: 'Densities of up to sixty persons per acre (net) are perfectly practicable with two-storey terrace housing and modest gardens. Densities up to ninety persons per acre (net) can be used to provide a good variety of housing types–some two- and three-storey houses, and some three- or four-storey flats–in schemes covering several acres. Tall blocks of flats will become necessary at densities higher than this but they need not predominate until densities of at least 140 persons per acre are reached.'[1]

The case for using less tall buildings becomes even stronger when one considers the relative costs of dwellings in buildings of different heights. Great play is often made by garden city enthusiasts with the relative costs of two-storey houses compared with flats, whereas what is really being compared is building *high* or building *low*. It should be possible to build two-storey flats–a flat per floor–more cheaply than two-storey houses if only because of the tighter circulation space and thus the smaller areas possible for a flat compared with a house. With three-storey flats, it is of course much cheaper to provide the equivalent smaller flats–one-bedroomed and bedsitter type of flats, than to provide small houses. It would even appear, if full advantage

[1] *Residential Areas: Higher Densities*, Ministry of Housing and Local Government, 1962.

of large-scale building methods were to be taken, to be as cheap to build two- and three-bedroomed flats in three-storey blocks, when the additional costs of roads, sewers and extra land are fully taken into account, as it is to build houses for the same family sizes. Four-storey maisonettes, with one maisonette having direct access to ground level and with staircases to the third floor serving the upper level of the maisonettes, are also almost as economical to build as two-storey houses, providing the same accommodation. Again, when everything is taken into account, including roads, sewers, work around the outside of the buildings and extra land required for lower densities, the same number of dwellings could probably be provided as cheaply in four-storey maisonettes as in two-storey houses. Eventually, the economics of prefabricated building systems used on a large scale to gain all the cost advantages of long production runs, could make walk-up flats and maisonettes at least as cheap to build as two- and three-storey houses. Experience in other European countries has already demonstrated this.

When high buildings are compared with these lower ones it becomes obvious that there are inherent costs which make it impossible for them to be cheaper. The lifts, staircases and access corridors, services and fire protection required add appreciably to the cost of each dwelling unit. It would appear that, when the costs of roads, sewers and external work are taken into account, 'building high'—with lifts—is often as much as one and a half times the equivalent cost of houses providing the same accommodation and standards.

It is a pity that these undoubted facts are often misinterpreted to produce a case against ever using tall residential buildings. Despite all the fulminations against living in flats there is evidence of a growing demand for them from people without small children, those whose children are older or whose children have married and left home as well as for old people who are still active. This question of living in flats is also often mixed in with an idyllic picture of everybody wanting their own garden. We all want usable outdoor space to sit and relax in. Children need play spaces of different types for different age groupings. Some of us, in addition, want gardens because we like gardening. But this fetish of every house with an ample garden is carrying the idealized version of rural 'Merrie England' to absurdity. About 60 per cent of Britain's population consists of single people, old people, childless couples and families with children over sixteen years of age. To fail to provide for a growing proportion of flats is to fail to provide adequately for these smaller households who often want, particularly in middle or old age, a less time and energy-consuming form of accommodation.

HOUSES FOR TOWNS

Housing layouts will need to change radically during the next decade, partly because higher standards of living affect house plans but mainly because of increased car ownership. The chief problems which new layouts will have to meet include the need to raise the densities of schemes with two- and three-storey houses to ranges of between fifty and eighty persons per acre. The increase is in order to economize in land, in roads and services, to avoid housing sprawl and to reduce walking distances between schools, shops and homes. Furthermore, there is the need to provide space for motor cars at a ratio of one car to each dwelling with additional parking space for visitors' cars and access for service vehicles. Not only is it necessary to separate pedestrians from motor traffic, but some local open space as an amenity within each residential unit is required as a proper setting for buildings, and a place where small children can play in safety or people can sit around and talk with neighbours or friends. Lastly it is important to improve privacy far beyond the level which is normally provided by the open back garden. Every house must have its own private 'outdoor room'. Already there are sufficient examples to show that all this can be achieved successfully, providing both a good environment as well as interesting vigorous architecture.

In an effort to consider the application of these needs to new forms of housing layout and new house types, the Ministry of Housing set up, in 1960, a Research and Development Group.

They have evolved house plans giving the larger amount of space within each dwelling and the increased flexibility and higher standards called for by the Parker Morris Report.[1] Designs put forward had at least two separate living spaces on the ground floor and flexibility in the arrangement of bedroom partitions to meet the changes which normally occur in the life of a family. One design for a five-person terrace house shows very clearly this re-arrangement of the living space to provide two good sized living areas separate from the kitchen as well as the addition of a small study-bedroom on the ground floor. By the elimination of one partition between the first and the third bedrooms, this design can easily be changed into a spacious home for four persons. The small garden is taken over to the same side as the pedestrian pathway systems where it can flow more naturally into the community open space beyond the garden gate.

Another design, on similar lines, is the three-storey house with the garage incorporated on the ground floor (Fig. 32). Again, this has been tried out with obvious success in the new towns, even though within tighter space allocations. This house type is adaptable to a wide variation of requirements, especially for larger families. Three-storey houses have architectural merit in making it possible to introduce variety in the skyline, to create points of local emphasis and to help introduce a more urban scale.

Developing from earlier examples is the patio-terrace house plan (Fig. 33). Here there are virtually three living spaces on the ground floor—the single-storey living room, the dining area which adjoins the kitchen and the study-bedroom. The new floor space standards adopted allow for more generous entrance lobby and storage and for a downstairs w.c. and wash basin; these are provided, as recommended by the Parker Morris Report, for five-person dwellings on more than one floor. Upstairs there can either be two double bedrooms or one of these can be subdivided so that when children are small they can be on the same floor as

[1] *Homes for Today and Tomorrow*, Parker Morris Report, H.M.S.O., 1961.

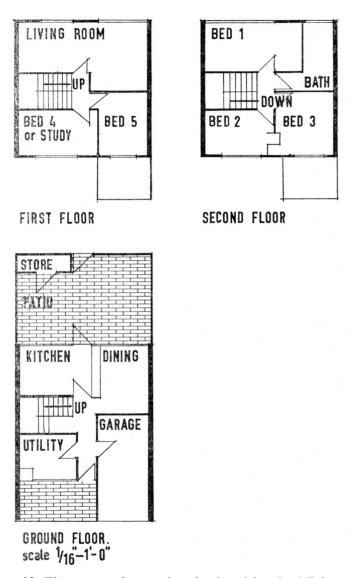

Figure 32. Three-storey house plan developed by the Ministry of Housing's Development Group, adaptable to a wide variation of requirements, especially for larger families.

24. Air view of the famous Roehampton Scheme. The flats, maisonettes and terraces of houses are powerfully grouped in a mature
landscape setting.

25. In the heart of the Roehampton Scheme; a group of single-storey old persons dwellings along a footpath with a backcloth of fine trees and elegant tower blocks.

26. General view of the West Ham pilot project of the Ministry of Housing's Development Group using new house types to Parker Morris standards (see Figs. 32 and 33). This view shows the limitations imposed by the relationship of the housing area to a major trunk road.

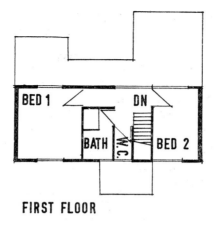

FIRST FLOOR

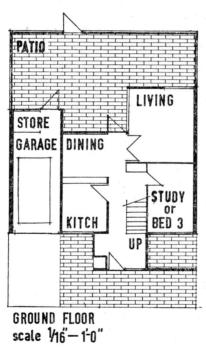

GROUND FLOOR
scale 1/16" — 1'0"

Figure 33. Patio-terrace house plan developed by the Ministry of Housing's Development Group. Three living spaces are provided on the ground floor.

149

their parents; when the children grow up, two of them can have a small room each and the eldest use the study-bedroom on the ground floor.

A pilot scheme using these house types was built at West Ham[1] on a small site of two and a half acres (see Plate 26). The density of the housing was approximately eighty persons per acre. After the scheme was completed a full-scale assessment was carried out. This consisted of an architect's assessment of the interior layout and construction, helped by diaries kept by a number of housewives of their activities in the home, together with a sociological survey of interviews with housewives. Not only was the technical performance of the house studied, by measuring heating standards achieved, sound insulation and daylighting, but a study of the use of the open space in the centre of the scheme was made by the use of time-lapse photography. It was found[2] that some of the private gardens which were less than 300 square feet were inadequate for sitting out, for younger children to play and for drying clothes. Also the low fencing provided between private gardens and the half-acre communal green gave insufficient privacy to the dwellings surrounding the green. Some of the faults detected reflect on the location of the scheme and its relationship with the surrounding area. A busy road to the nearby docks separated the scheme from the primary school and the nearest park. Consequently, the small open green became too intensively used, especially by older children who, it had been assumed, would play in the park instead. This pilot scheme should have shown the difficulties of achieving a satisfactory environment by patchwork methods; many of the faults that came to light could be overcome in schemes of larger dimensions where housing is placed in a proper relationship with open space, schools and shops. But this scheme also demonstrated how effectively houses at higher densities could be provided in towns, catering for modern needs and including a high proportion of garages and parking spaces within the scheme.

[1] Now part of the London Borough of Newham.
[2] See article in *Architectural Design*, August, 1966, on the work of Research and Development Group.

RENEWING TOWNS AND CITIES

The main examples of city and town renewal can be grouped into three categories. First, there have been a number of outstanding large-scale schemes on the outskirts of cities and larger towns which have broken away from the more characteristic suburban pattern and constitute small townships within the city; even so, they are really forms of peripheral expansion. Perhaps the best known early example of this type is the London County Council's Roehampton scheme. Secondly, there have been a number of comprehensive schemes which are predominantly for new residential areas located near the centres of cities and towns but which have involved the more complex problems of the replacement of old buildings by new. Thirdly, there have been the beginnings of the renewing of older city and town centres themselves; this latter aspect forms the subject of the subsequent chapter. It is useful to discuss the more successful approaches which have been made to urban renewal in this order as this, more or less, is the chronological pattern – the peripheral schemes were under way first, central redevelopment housing schemes second and the central areas (the most complex problem of all) – only just beginning to be seen in their built form.

Of course, a great deal of housing development in the outskirts of cities has not shown any substantial differences from schemes developed before the Second World War. In particular, many private and public development schemes have almost completely disregarded proper provision for the motor car. Reference has already been made to pioneering attempts made during the 1920's and 1930's in the United States to come to grips with this problem of the motor car in residential areas. We have discussed the origins of the Radburn-type layouts where the car is more adequately catered for and where separation of pedestrians and vehicles is the main aim. In Britain, the Radburn idea was comparatively slow to make a start after the war. It first made its main appearance at Wrexham in 1950 in a scheme prepared by Gordon Stephenson. Other schemes followed at Northampton, Stourport and Sheffield – but these were isolated

examples. These schemes were discussed in some detail in an
article by Gordon Stephenson in 1954.[1] It is of interest to see how
the main emphasis, at this time, was placed on the effects of this
new type of layout on *house costs*, rather than on the provision
for a large percentage of car ownership or for separating the
pedestrian from the vehicles on the grounds of safety. Looking at
these early Radburn layouts it emerges that the provision of
garages and parking spaces was totally inadequate. But the fault
does not necessarily lie with these pioneer designers who had
other difficulties to contend with. Thus Stephenson comments in
the article referred to above: 'Though there is a demand for
garages the Ministry have ruled that none shall be built until it is
presented with a list of signatures of those requiring garages. In
some places the garages are intended as screens and are part of
the composition. It is more expensive to build them at a later
stage and, apart from this, garages are buildings on which the
local authority can, with justification, make a profit.'

With all these earlier layouts there was a strong degree of
compromise in the separation of the footpath and road systems;
footpaths often led to busy roads which had to be crossed to
reach shops or schools, or footpaths were provided on either side
of a road despite the fact that there was to be a separate footpath
system. These schemes were also usually only small parts of
larger developments which had little claim to be comprehensive;
the full flavour of the Radburn layout, with its emphasis on complete separation of road and footpath systems and a degree of
self-containedness for a large housing area, was still proving to
be illusive in its achievement in Britain.

A major step forward in applying Radburn layout on a larger
scale was achieved in the Willenhall Wood schemes at Coventry
(Fig. 34). Three schemes have now been completed, totalling
more than 1,100 dwellings and including shops, a nursery school,
public house and other community buildings. The first of these
schemes was completed by the end of 1958. Provision for garages
was much higher than for many previous public housing schemes
in Britain–about one garage to every two dwellings–and the

[1] Gordon Stephenson, *Town Planning Review*, January, 1954.

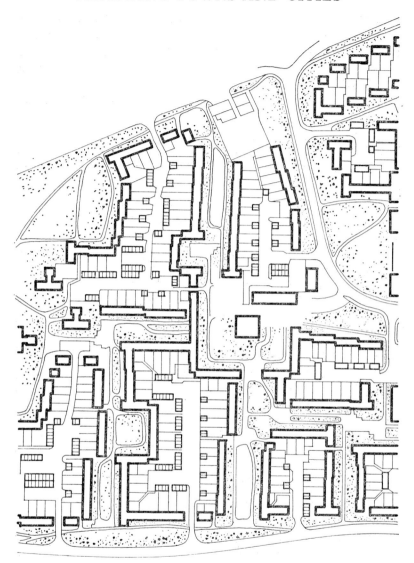

Figure 34. Willenhall Wood, Coventry. Part of the first stage of a now well-known scheme on Radburn lines which was completed at the end of 1958. Garaging for about 50 per cent of the dwellings has been provided.

footpath system was very much more comprehensive. The scheme showed that a pleasant environment could be created by utilizing Radburn layout on a much larger scale, providing a high proportion of garages and parking space and also reaching a relatively high density for two-storey houses and flats–about seventy persons per acre. However, this first stage of the overall scheme revealed other problems which had then to be tackled. There were long stretches of garages to the rear of houses with low fences which denied the possibility of increased privacy which these layouts set out to provide. Rear access to the dwellings was monotonous and dreary (see Plate 28). Moreover, the classic Radburn idea assumed that, wherever there were busy roads, these could be crossed by footpaths either underneath or by bridges over the road; the need for pedestrian underpasses or overpasses was a fundamental part of the Radburn concept. These were lacking at Willenhall Wood as indeed they were in all the earlier types of Radburn layouts, both in the new towns and in city and town renewal schemes later. In the second stage of Willenhall Wood improvements were made in the garage courts, with higher fences and garages in smaller groupings. The third and final stage of the scheme, completed in 1966, provides garaging in courtyard areas with access through openings in the terraces of houses thus overcoming some of the visual problems of the garage courts in the earlier stages. There are also further improvements in privacy. The scale of provision of garages in the third stage has now become one per dwelling together with very substantial additional car parking for visitors. Pedestrian underpasses are now being provided at Coventry on later housing estates to reach schools, open spaces or other parts of the residential areas, even though the roads concerned are estate roads and not principal traffic routes. The way in which improvements have been made stage by stage through the years that these schemes have been in construction and the way in which lessons are applied to later schemes is an excellent example of how a progressive authority should work.

The Radburn concept took a further step forward with the

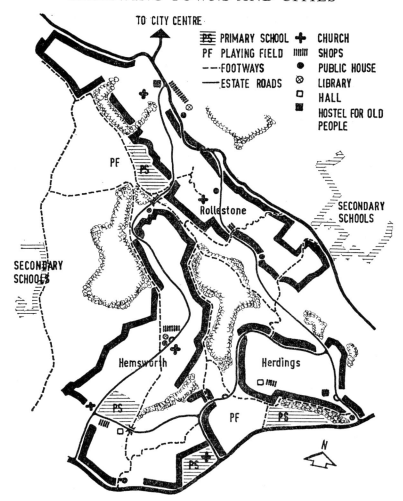

TO CITY CENTRE

PS PRIMARY SCHOOL   ✚ CHURCH
PF PLAYING FIELD   ⅢⅢ SHOPS
-- FOOTWAYS   ● PUBLIC HOUSE
— ESTATE ROADS   ⊗ LIBRARY
  □ HALL
  ▣ HOSTEL FOR OLD PEOPLE

PF

PS

Rollestone

SECONDARY SCHOOLS

SECONDARY SCHOOLS

Hemsworth

Herdings

PS

PF

PS

PS

N

Figure 35. Gleadless township, Sheffield. Three neighbourhoods totalling more than 17,000 people with separated pedestrian and road systems.

scheme for Sheffield at Gleadless Valley (Fig. 35). Here, a whole new township of just over 17,000 was developed on Radburn lines. There are three neighbourhoods each with its own primary school and shopping centre. The neighbourhoods are naturally

155

defined by the rolling nature of the country and the names of the three neighbourhoods have local associations, Hemsworth, Herdings and Rollestone. The nearest of these three neighbourhoods is to be a district shopping and social centre to serve the three neighbourhoods and also adjacent pre-war estates. A footpath system has been planned independently of the roads linking together the three neighbourhoods and linking the dwellings in each neighbourhood with its schools and shops. The whole site occupies a total of 450 acres of which 161 acres are to be devoted to parkland, woodland and playing fields. The site is steeply

Figure 36. Houses down the slope at Gleadless, Sheffield. A garage basement for the houses on one side of road has the space for two cars, sufficient to serve the terrace of dwellings lower down the slope.

sloping ground and imaginative use has been made of the different levels to enable as many houses as possible to have the best aspect for sunlight and a view and to use taller blocks of flats to greater architectural effect. The residential density of this township is seventy persons per acre. In addition, the sloping nature of the site has been used to great advantage to provide garages and parking in the slopes underneath dwellings so as to make them less obtrusive (Fig. 36 and Plate 29). Underpasses have also been included in this scheme, not as many perhaps as are really needed, but nevertheless linking housing areas with schools. The whole scheme is bold architectural and town planning and a model of what could be done in many other suburban areas of cities and towns.

A number of other schemes which have been carried out on the outskirts of cities are of note because they are not just

27. Housing at Willenhall Wood, Coventry, with separation of pedestrian footways from vehicle roads. The scheme was completed in 1960 (see similar layout on page 153).

28. The 'backs' of the same scheme. Garaging is provided for half the dwellings. The dreary garage courts of early 'Radburn-type' layouts are their main weakness.

29. Double garages under three-storey houses provide also for split-level houses on the opposite side of the road (see diagram on page 156). Housing at Gleadless Valley, Sheffield.

30. A neat, unobtrusive solution to garaging in two storeys at the Calthorpe Estate, Edgbaston, Birmingham.

peripheral housing but substantially new types of *urban* development. The traditional 'pyramid' of densities–whereby highest densities are used for housing towards the centre of cities and densities become lower the nearer one reaches the edges of the city–have been disregarded. Thus the Roehampton scheme of the former London County Council is really a new township in the suburbs and thus makes a contribution to the needs of housing for Londoners from the more crowded areas. It makes its contribution by giving vital manoeuvring space in central and inner areas of towns and cities without adding to already gigantic requirements for dispersal of population to new and expanded towns outside.

Up to the time when it was completed, in the late 1950's, Roehampton was one of the largest housing schemes in Europe (see Plate 24). It caters for a total of 9,500 people and the overall residential density is about 100 persons per acre. The site of 130 acres consists of the grounds of large eighteenth-century mansions. Many buildings other than dwellings have been provided–sixteen shops, two club rooms, three primary schools and a secondary school will adjoin the area. The original mansions themselves are also used for community and social purposes. The scheme is related to the old village centre of Roehampton and therefore derives considerable benefit from established shopping and other facilities. The concept of 'mixed development' has been employed, using a mixture of both houses and flats to excellent advantage. The scheme is noteworthy because of its use, virtually for the first time in Britain on a large scale, of tall, residential tower blocks. There are twenty-five such eleven-storey tower blocks in all and the whole project includes twelve-storey maisonette slab blocks, terraces of four-storey maisonettes and terraced houses–all of considerable architectural merit. The buildings are set in a wonderfully mature landscape in the best traditions of eighteenth-century town planning with a careful, studied but informal grouping. However, as far as garaging and the segregation of pedestrians and vehicles were concerned, Roehampton did not make a significant contribution; it may be possible to inject more facilities for the motor vehicle at this later stage but this can be no substitute for a layout which starts from the conception of separate footway and vehicle paths.

Of the considerable amount of housing by private developers and builders that has taken place since the war, there is very little which points the way forward by trying out new forms of layout. Nearly all post-war private development housing has followed the familiar patterns of the 1930's and relied on roads in front of all houses to provide the means of access both for cars and pedestrians. Of the few significant examples of pioneering private development, the most notable have been the schemes, designed by the architect Eric Lyons, for the firm of Span Developments. Catering for a relatively specialized market, 'Span' has provided dwellings–flats and houses–which are almost entirely for the executive, professional and middle-class income groups. This has meant that venues for Span Developments have had to be very carefully chosen; the normal sites still available in the outer suburbs have tended to be ruled out by the middle-class attitude of potential residents. The most fashionable London suburbs also tend to become ruled out by the very high property values now prevailing there. Thus the main sites for Span Developments tend to be the older but more arcadian suburbs around London–Richmond, Kingston, Teddington and Twickenham, Blackheath and Beckenham. Additionally there have been some signs of this sort of development spreading to some of the provincial towns, such as Cambridge (Fig. 37) and Hove. In its early stages, Span Developments was not outstanding for its use of bold, new ideas in housing layout. The segregation of pedestrian and vehicle remained very limited. The small scale of many of these developments did not enable them to tackle the more ambitious pedestrian footpath systems distinctive in some of the larger public authority schemes already referred to. Of the earlier 'Span' schemes, Parkleys at Ham Common has a considerable degree of segregation with garages on the periphery of the sites, but for the 166 flats only sixty-six garages were provided (Plate 31).

A similar type of development has been undertaken by the Calthorpe Estate at Edgbaston, Birmingham. Here, the ownership of a very large area of land by one estate company has given great opportunities for overall layout. The whole of the Calthorpe Estate, which is more than 1,600 acres in extent, is under the leasehold control of the estate company. This has enabled the company to have a master plan for the whole area prepared by John Madin and Partners, the estate's architects; but the fact that

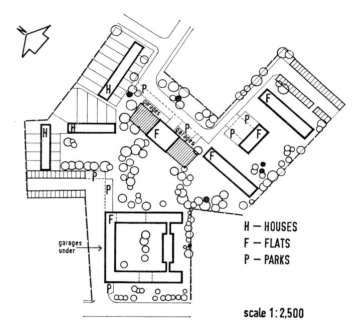

H — HOUSES
F — FLATS
P — PARKS

scale 1:2,500

Figure 37. Layout of flats and houses for Span Developments Ltd. at Cambridge. Architect: Eric Lyons.

it is necessary to wait for leases to fall in means that the pace of development tends to be very slow and the development can only proceed in a piecemeal fashion. Nevertheless, the concepts of segregation of pedestrian and vehicle are being taken a great deal further here than in many comparable private development schemes. Also new ideas in the grouping of garages and the provision of garaging on more than one level have been tried out (see Plate 30).

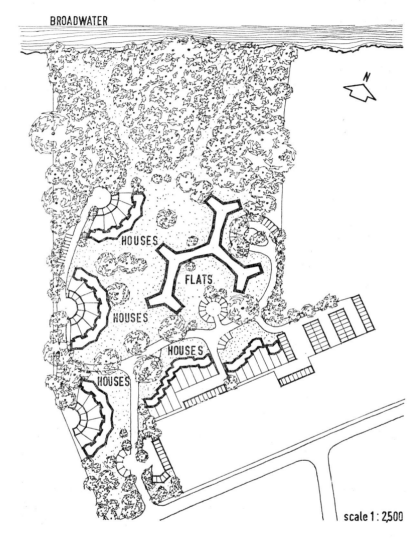

Figure 38. Private houses and flats for sale. One garage per dwelling and plentiful parking space for visitors. Scheme by architect Eric Lyons at Weybridge, Surrey (Plate 32).

31. Flats for sale. An early scheme at Ham Common, Surrey, which started a movement among private developers.

32. A scheme of houses and flats at Weybridge, Surrey, for the same developers. Garaging is grouped mainly around the outside of the area leaving the centre for footpaths between the trees (see layout on page 160).

33. A model of part of the Greater London Council's riverside development at Woolwich/Erith where open space and water will provide an attractive environment in a riverside town for 60,000 people.

34. The Barbican scheme. Intricate planning has made it possible to inter-mix flats with offices, shops, cultural and educational buildings. The whole scheme is planned at several main levels with garaging at ground and basement levels for 2,600 cars.

RENEWING TOWNS AND CITIES

One of the most interesting examples of new layout forms and
provision for the motor car in a private development scheme is
the scheme by Eric Lyons for 'Span' at Weybridge (Plate 32 and
Fig. 38). Here, two long culs-de-sac run either side of the site
serving most of the garages on Radburn lines; almost complete
pedestrian and vehicle separation is achieved with the maximum
retention of mature trees. There is one garage per dwelling with
66 per cent more parking space for visitors in addition. As with
so many of the schemes by Eric Lyons, great use is made of
existing landscaping. The site has some superb trees and the
scheme has been designed to exploit the scenic qualities of the
landscaping with the shaping of the buildings in flowing crescents
devised to create long vistas and establish a spacious environment even though the density is higher than most comparable
private development housing—about sixty-four persons to the
acre.

RENEWAL IN LONDON

Of urban renewal schemes in the inner areas of British cities,
the first in importance is undoubtedly the area, started by the
former London County Council, in Stepney and Poplar. This
scheme has now been inherited by the Greater London Council
who are the present planning authority for this part of the London
Borough of Tower Hamlets. The Stepney/Poplar reconstruction
area, as it was originally called, was far greater in its compre-
hensiveness than anything else being attempted in the early
reconstruction period in Britain. In addition, because of its
relatively early start, there is now a great deal more to be seen of
this scheme on the ground. The whole area is almost 2,000 acres
and is to provide eventually for a population of nearly 100,000
people. It should perhaps be mentioned that, in the same area,
there were 217,000 people living in 1939—before wartime evacua-
tion, bombing and destruction reduced the population. It was
the large-scale clearance of huge areas which undoubtedly
facilitated an early start to renewal in this part of London. The
whole scheme will take at least till the mid-1970's to complete

and is, therefore, only something like half-finished to date. It provides twelve neighbourhoods varying in size from 2,200 to 10,700 people (see Fig. 39). The neighbourhood structure which has been used has been a flexible one relating, as far as possible, to the existing communities already living in this area. The work on the scheme was divided, by the London County Council, into several periods. In their *London Plan, First Review–1960*, it was stated by the L.C.C. that in the first period of this development–

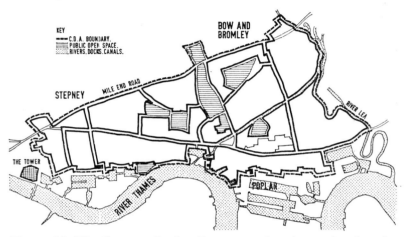

Figure 39. The Stepney–Poplar Reconstruction Area, showing the sub-division of the area into neighbourhoods of varying size and population.

from 1955 to 1960–although they had planned for development of 221 acres, only 144 acres were developed–77 acres short of the planned proposals. During this first period of the plan it was also proposed that there should be seven new schools and 20 acres were allocated for this. But it did not prove possible to build any of them in this first stage, although in many cases the land had been acquired and cleared. There were also lags in the provision of shops, landscaping and parks. Therefore, the L.C.C. drew the lesson: 'The real planning achievement is only realized when a whole new neighbourhood takes shape. Although good progress has been made in new housing and in land acquisition, development has been unbalanced. The new housing estates have not yet their proper complement of the new schools and parks which were programmed in the plan for simultaneous construction. . . . If the full benefit of a new environment is to be realized in schemes of comprehensive development, then it is necessary to construct the new housing, roads, schools, parks and shops simultaneously.'[1]

This was the initial period of this great scheme. The next period was called by the L.C.C. the 'Seven Year Plan'–from 1961–1967–and this aimed to complete seven of the twelve neighbourhoods including 40 acres of new parks, thirteen new schools and the construction of two major shopping centres and a new commercial and industrial centre near Aldergate. Although progress has continued to be disappointingly slow, it is hoped that several neighbourhoods will be completed by the end of the 1960's. Nevertheless the total achievement is impressive. Since the end of the Second World War, 13,000 new dwellings have been built, four secondary schools and seven primary schools completed and 39 acres of new public open space have been laid out. So that, although there are weaknesses in carrying to fruition this ambitious project, its immense achievements are paramount. The finished scheme will have brought about the renewal of a large slice of London, catering for industry and commerce as well as for the houses, schools and community buildings which go to make up London's East End.

However, it must be said that the rate of development has been far too slow. For decades its inhabitants must live in the midst of obsolescence gradually overtaken by demolition, the transformation of large parts of the area into great building sites and then, gradually, new buildings; for local residents there must seem no apparent end to an endless process of construction going on continually around them. Within the Stepney/Poplar area, the complexity of different uses and the difficult and costly process of land acquisition (at the inflated values which the development itself has tended to produce) have been formidable obstacles to faster progress. Moreover, the removal of obviously unsuitable

[1] London County Council, *London Plan, First Review–1960.*

industry has only been able to be accomplished by 'buying them out'; and the money available for this has been insufficient to tackle any but the worst non-mixing industry. The main commercial and shopping streets–Mile End Road and Commercial Road–remain almost untouched by the redevelopment going on behind them. Above all, this scheme did not attempt the difficult task of securing complete separation of pedestrian paths and roads for vehicles within each of the neighbourhoods. Although there has been a drastic limitation of through traffic–discouraging it from going through neighbourhoods–and road patterns have been simplified with fewer outlets to main roads and systems of pedestrian footpaths provided, the whole scheme will inevitably suffer because of the penetration of the motor vehicle within neighbourhoods. Unfortunately, also, as with so many other areas started immediately after the war, the provision of sufficient garaging was not seriously tackled. This is bound to lead to severe limitations to the ultimate success of the whole project, even though attempts are now being made to remedy the deficiency wherever possible.

Another outstanding urban renewal project is the Barbican Scheme. The difference between this scheme and the Stepney/ Poplar scheme is, for the most part, the difference between the relative affluence of the City of London compared with that of the East End of London. The Barbican Development is outstanding because of the boldness with which the problem of the motor vehicle is tackled, with separation of vehicles and pedestrians at different levels, and the way in which intricate planning has made it possible to inter-mix residential accommodation with commercial, shopping, cultural and educational uses in an imaginative and exciting way. The scheme attacks the roots of the concept of the City of London as only a business and commercial centre which 'dies' after working hours. A planned admixture of dwellings and other compatible uses can produce a new form of urban life with its own community and social life, attracting back those people who find life in a truly urban environment to be stimulating and who can afford the relative luxury of living there.

RENEWING TOWNS AND CITIES

The Barbican Scheme really consists of one area, mainly residential, covering an area of 40 acres which is being closely linked with a business and commercial area to the south, a further 22 acres in extent. The commercial and office area has not only tall office blocks but also shops, pubs, restaurants, warehousing, a museum and an underground station. The residential area will provide new living accommodation for 6,500 people in some 2,100 flats and houses. The whole scheme has been designed to a very high residential density, originally planned at 300 persons to the acre, but now reduced to 230 persons to the acre. There are to be three, forty-four-storey, tower blocks. In addition to dwellings, the area will contain schools, an art centre, shops and recreation facilities. The whole scheme is planned on several levels with garaging at ground and basement levels and with the upper levels above the garages laid out as terraces and gardens. The intention is to provide, for the larger flats, one covered parking space per flat and, for the smaller flats, a covered parking space for 50 per cent of the flats. There is also additional garaging for a further 500 cars from the commercial area to prevent street parking in that area; this additional garaging will also serve for additional parking in the evenings and at weekends.

It should be mentioned that the Barbican Scheme, also, was extremely slow to start. It has taken something like twenty years of protracted negotiations before it was possible for the bulk of the scheme to begin and it will not be completed until the early 1970's. However, an initial project at Golden Lane to the north which was designed by the same architects as the Barbican Scheme–Chamberlin Powell and Bon–is now completed and gives a preview on a smaller scale of the quality to be expected of the larger scheme. The 7-acre site at Golden Lane provides residential living for 1,400 people in flats and maisonettes at a relatively high density–200 persons per acre (Fig. 40). But the scheme is noteworthy for the range of facilities included apart from dwellings. These include tennis and badminton courts, swimming pool, laundry, creche, youth club and senior citizens club. The community building has a meeting hall, committee room, kitchen and recreation rooms; there are also seventeen shops, two restaurants and a pub in the housing area. The site was planned at two levels with substantial terraces and gardens;

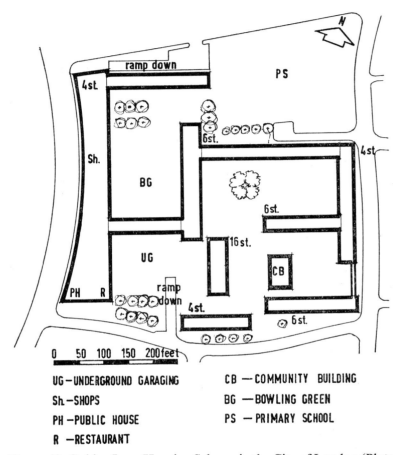

Figure 40. Golden Lane Housing Scheme in the City of London (Plate 37). A forerunner of the Barbican project (Plate 34).

the old basements from this original bombed site were utilized to provide garaging and a basement road servicing the shops in the scheme (see Plate 37).

The rents of the flats to be provided in the Barbican Scheme, as with the Golden Lane scheme, are to be high. Thus a two roomed flat will probably be at least £500 per year and a four-roomed flat in a tower block at least £850 to £1,000 per year, these figures being inclusive of rates and heating. Such rents have led to criticisms because accommodation for lower income groups has not been provided. Yet the scheme marks a victory for planning and comprehensive redevelopment over piecemeal and small-scale schemes. Because it aims to provide residential accommodation mainly for middle or even higher income groups, its lessons are not likely to be universally applicable. But the bold, imaginative sweep of the scheme shows what can be done where large numbers of people are to be provided for in a highly urban environment. It is not so much that this scheme provides standards of accommodation and living environment which are much higher than the normal, although it will certainly do this, but that it takes all the elements of a well-equipped residential environment and disposes them over the site extremely well. Any scheme attempting to produce attractive urban conditions like these, including a wide range of facilities to help entice people to live in the centre of cities, is bound to be expensive. While such wide disparities of income exist, it is surely not unreasonable to provide such relatively costly accommodation for those who prefer to live in the heart of cities rather than commute to the outskirts.

STREETS IN THE AIR

A further outstanding example of urban renewal in practice is Sheffield's large Park Hill scheme, together with the later Hyde Park scheme which is closely associated with it. Here is an attempt to provide high density housing in a new form with pedestrian circulation separated from motor vehicles not only at ground level but by 'streets in the air'. The Park Hill scheme was completed in 1961. Eventually the two schemes will provide dwellings for some 8,000 people at a density of about 200 persons to the acre.

The main reasons for the selection of this particular site for a high density scheme appears to have been the fact that this area was already cleared before the war and included some of the oldest and worst slum property still standing. By building to a high density on an area where comparatively fewer houses had to be demolished additional accommodation could be provided. In the area as a whole only some 800 dwellings had to be demolished but 2,307 dwellings were provided in the scheme, giving a surplus of about 1,500 dwellings to enable the city to continue its clearance of slums elsewhere.

The site rises sharply to about 200 feet above the Sheffield Midland main line railway station and the industry in the valley. It was considered to be a suitable site for high density housing despite its close proximity to heavy industry in the valley below; the prevailing winds, it is claimed, blow industrial smoke and pollution away from the scheme and the high flats have all the advantages of better air, light and views as well as some open space nearby.

The principle adopted in the layout (see Fig. 41) is of vehicular and pedestrian separation. The through roads are kept at the edges of the housing area and not allowed to cross through it. The primary school and playgrounds are lodged in the centre of the scheme. Vehicular roads are short culs-de-sac. There are 100 parking spaces for private cars provided in the first part of the scheme (995 dwellings) and these are mainly intended for visitors; there are also lock-up garages, the roofs of which are to be used as children's playgrounds. It is proposed that, when demand for garages increases, a multi-storey garage with circular ramps should be added and provision for this was made in the plan.

The main feature of this scheme is the pedestrian street decks, 10 feet wide, open to the air and provided at every third floor, bridging across between the blocks. Thus the buildings have a continuous street system around the whole site on every third floor (see Plate 35). The deck system is served by lifts for passengers and large goods lifts which are used for electrical milk floats, furniture removals, etc. As the site slopes along its length and the roof line is maintained horizontally, the blocks vary from fourteen storeys at the highest end to four storeys at the lowest.

Thus all the horizontal decks, except the top one, run out at ground level at one end. This helps the horizontal circulation by reducing the number of access points needed at ground level and simplifying the footpath system required; it also achieves considerable economy in the provision of lifts.

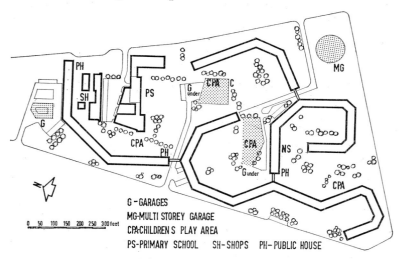

Figure 41. Park Hill, Sheffield. High density housing based on 'streets in the air'—a continuous deck system within the buildings at every third floor. At ground level, cul-de-sac roads lead to garages and serve shops, pubs and primary schools without cutting the site into separate parts.

The promenade decks take the place of streets, providing places for social, face-to-face contact. Within the dwelling there is privacy but the deck is spacious enough for neighbours to stand and chat, for the milkman's float and even for children to play to a limited extent. A further feature of the scheme is the way in which other amenities have been included. There are shops, a community hall, four pubs, both primary and nursery schools, laundry and central heating plant.

One may admire the vigour and the boldness of the concept of the Park Hill scheme but still question whether it is entirely valid as a solution from a town planning point of view. There is little

169

doubt that the whole scheme looks visually exciting, especially when viewed from within or at night from the city centre. But one cannot help feeling that it is an attempt to solve Sheffield's housing problem by perpetuating high density living on the wrong site. No one who goes and looks at the housing and town planning problems of Sheffield can doubt the difficulties. The slopes of the ground, the outworn slums, the intermixture of

Figure. 42. A pedestrian deck in a large housing scheme for the former Croydon Airport, by Clifford Culpin and Partners, showing the garages at ground floor level and the ingenious interlocking houses.

noisy industry are all serious problems. But one cannot help asking whether the answer should not lie partly in more separation of housing from heavy and dirty industry, not necessarily within Sheffield's boundaries, but by a new town in South Yorkshire or Derbyshire?[1] The Park Hill concept has great merit in itself, but in its particular environment–as a cliff of buildings overhanging the railways and noisome industry–the whole project has a giantism which leads to doubts. Nevertheless, the concept of 'streets in the air', and the use of levels to provide servicing to the blocks without interfering with pedestrian circulation, are principles which appear to add a considerable amount to our thinking on these questions.

BACK TO BATH

Since Park Hill, there have been other projects developed which appear to take the main ideas further. In the study of

[1] The extension of Sheffield's boundaries to include a new satellite town for 80,000 people at Mosbrough is now under way. This solution was not available when Park Hill was conceived.

Fulham in London, undertaken for the Ministry of Housing by
the Taylor Woodrow Group, the possibilities of long, continuous

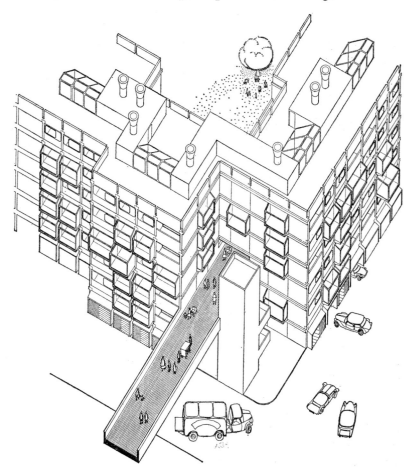

Figure 43. Fulham project: a multi-storey block with a pedestrian
deck and vehicle circulation and car parking below.

blocks with 'street decks' were also proposed. A six-storey
continuous block form was developed, with a continuous street
deck through the block at second floor level (see Fig. 43). Alternative layouts were prepared to densities of 136, 200 and 250

persons to the acre with resulting variations in the grouping of blocks around open spaces. Garaging was tucked under the six-storey buildings. Comparisons were made in the scale of this continuous, six-storey block and Georgian planning in London's Bloomsbury. The height of these blocks and their disposition into squares accorded very closely with Georgian terraces of four-storey dwellings and could, it was claimed, give the same sense of city scale which is required of urban building. Unfortunately the Fulham study was not proceeded with; it remained a theoretical pilot study.

But similar concepts have motivated a large scheme in the Hulme area of Manchester prepared by Hugh Wilson and Lewis Womersley (who was city architect of Sheffield at the time of Park Hill). This scheme was under consideration by the Manchester City Corporation in 1967. The task was to provide housing for about 1,000 households, varying from two to nine persons in size, at a residential density of about 100 persons to the acre. The housing was to be association with the Hulme neighbourhood centre which will serve 15,000 people. The designers studied the great terraces of Bath and Regents Park, London, and decided to make it their aim to create an environment of city scale by using four great curving six-storey blocks, each with continuous access decks on the lines of the Park Hill solution. Here is a conscious attempt to design a twentieth-century version of Georgian terrace housing (see Fig. 44). Pedestrian decks not only give a separation of pedestrians and vehicles but also free the ground for treatment as a landscaped 'park' with bold forms – grass banks and belts of trees – which shelter sunken play areas and partly hide roads from view as well as sheltering dwellings from the noise of traffic. The other important innovations compared with Park Hill are the way in which the pedestrian decks are to be linked with the neighbourhood centre, providing sheltered walkways; also the use of the continuous blocks for a more radical solution of garaging and parking problems; garages can be incorporated under the dwellings and served from an access road on the opposite side of the block away from living rooms.

Perhaps the most important difference between this scheme at Hulme and many of the other large housing schemes we have been considering is the way it fits into comprehensive planning, relating to the broader planning of urban renewal in Manchester as a whole. This is the first of four major inner areas of the city being redeveloped comprehensively and the detailed scheme

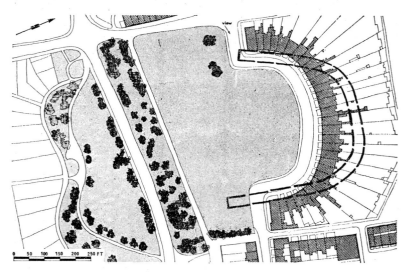

Figure 44. Plan of the Royal Crescent, Bath, with one of the six-storey continuous blocks of the Hulme scheme, Manchester superimposed, showing the similarity of size and relationship to open space.

described above has been designed within a 'planning brief' which set out the main planning principles to be followed. This has meant that the detailed scheme, of 38 acres, has had to be conceived as only a part of a very much larger area of 350 acres which forms the proposed Hulme neighbourhood of the future. And this neighbourhood, in turn, only forms part of the much larger city. The brief for Hulme, established by the City Planning Department, determines the relationship between this neighbourhood and the treatment of its centre part–the proposals described above–within the wider urban design framework. This means, for example, that it should be possible to avoid the detailed scheme being just another housing area with a series of tall blocks, unrelated to surrounding development and other building proposals which, unfortunately, is the piecemeal way in which so much of the urban redevelopment in British cities has taken place. The designers have had to take into account the fact that Hulme adjoins the education precinct of Manchester University and other higher education institutions–a great campus for 28,000 students which will include many groups of tall buildings leading to the even more powerful scale of the city centre itself. Hulme also adjoins other neighbourhoods and, in particular, must be related to the major district centre to be located nearby for the whole of the district of Moss Side which is to serve some 60,000 people and of which Hulme is only a part. The brief established for the Hulme neighbourhood was that it should be clearly definable as a residential area and that it should be a truly urban environment on a city scale but with its place in a wider pattern of surrounding development. Consequently, the brief stipulated that housing densities should increase near the neighbourhood centre. The Wilson and Womersley proposals, for higher densities but with medium height six-storey continuous blocks, aims to give that degree of distinct character required to the centre of the neighbourhood so that the detailed scheme fits in with the broader urban design concepts.

COVENTRY

The City of Coventry has also made great efforts to tackle the renewal of its inner areas, though increasing financial difficulties–the high cost of land acquisition and the difficulties of dealing with scattered areas of obsolescence rather than widespread slum clearance areas–have tended to make progress slower than the City Council would have liked. The problem of urban renewal in a city like Coventry was vividly portrayed by the City Architect and Planning Officer, who reported in 1959 that, of the 23,951 dwellings built in Coventry since the war, 23,773 had been built on the outskirts compared with only 178 dwellings provided in the inner areas near the centre. He commented: 'Coventry has not escaped the general trend of outward movement that has taken place in house building since the end of the war. It was obviously much quicker, simpler and cheaper to build new neighbourhoods on virgin soil on the outskirts of the city than to tackle immediately the slums in the older areas around the central core.'[1] Since then far greater efforts have been made to develop the inner zones but by the end of 1966, although Coventry's suburban housing has now reached more than 38,000 dwellings (of which 22,663 were private houses), the number of dwellings in the inner zone redevelopment areas was still only 815 and efforts to interest private enterprise in urban renewal housing had not been successful.

There are two central comprehensive development areas in Coventry–Hillfields and Spon End. The larger of these two areas is Hillfields and it is worth considering this as an example of the way in which Coventry has been tackling its urban renewal. When redevelopment started, the population of Hillfields was 5,600. The population will eventually reach 6,500 people–30 per cent in houses and 70 per cent in flats with an average residential density of 116 persons to the acre. There is to be provision for one car per dwelling. When renewal started, there were seventy-one factories in the area, ranging from a large factory for the Rootes group of vehicle manufacturers to small workshops. These smaller factories are now being regrouped into three defined areas around existing industrial concentrations. A main pedestrian route will link Hillfields with the city centre, passing through the area of high flats to the north-west of the scheme and then into the pedestrian shopping square, carrying on right through the area. Provision of garaging and parking varies in the scheme from a podium of several floors of garages at the base of the thirty-storey block (with the roof of the podium used for adults' and children's recreational facilities) to individual garages in the three-storey and patio houses (see Plate 38).

When it started the Hillfields area was only part of a wider

[1] Arthur Ling, *The Living Town*, Symposium of the Royal Institute of British Architects, 1959.

neighbourhood; as the report for the redevelopment stated: 'The boundaries of the Hillfields C.D.A. do not denote any social or human demarcation, but simply indicate the edges of a convenient "first bite" at the redevelopment problem'. Since then, however, the Hillfields area has been extended to natural boundaries related to standards of obsolescence and is now contained within a primary road system, thus creating a more logical neighbourhood. Originally, also, there was a main radial road of the city cutting through the area which meant the provision of underpasses to connect the two parts of the area together. The Review of the City Development Plan, however, provides for the deletion of this primary radial traffic route and realigned routes now form the boundary of the neighbourhood rather than passing through it. These two changes are good illustrations of dynamic planning in action, constantly seeking improvements in original concepts as schemes progress. Hillfields is now about halfway to completion and should provide a notable example of large-scale urban renewal near the centre of a city.

THE OBSTACLES

It is easy to be critical of the efforts that have been made, the beginnings of urban renewal in Britain's towns and cities. These efforts have to be seen against the background of almost crippling difficulties and obstacles. When the foundations of post-war planning policy were laid, in particular by the 1947 Town and Country Planning Act, local authorities were given very wide powers enabling them to acquire land and to meet claims for compensation out of charges on improving values of land arising out of the activities of the community. Even so it was assumed that redevelopment could often only be carried through at a loss and therefore exchequer grants were available towards the losses incurred on acquisition and clearing of approved sites. Local authorities also had considerable powers to undertake development on land they had acquired or to lease the land to private developers. It was assumed at that time that local authorities were to play a very direct role in redevelopment, taking the lead

35. 'Streets in the air' are the basis of the Park Hill scheme at Sheffield. These decks, ten feet wide, give continuous circulation around the whole scheme at every third floor, connected to passenger lifts and large goods lifts for the use of electric milk floats and for furniture removals.

36. Park Hill by night, showing the continuous decks at every third floor and the children's play spaces and garaging at the lower level.

37. Golden Lane housing scheme for the City of London Corporation. In front of the 16-storey flat block is the community hall with gardens at a lower level (see layout on page 166).

38. A glimpse of a future inner area of Coventry—the Hillfields Development Area. Covered pedestrian ways thread through tall blocks of flats towards the suburban shopping centre. Beyond are three-storey houses and patio houses. Crossings of main roads are by bridges or underpasses.

and setting the example which private developers could follow. The Government booklet *Town and Country Planning 1943–1951* explained: 'Because of the multiplicity of ownerships usually involved, the key to proper redevelopment of towns is public acquisition, to be followed either by the disposal of land to private developers, under conditions ensuring that they will themselves carry out development in accordance with the plan for the area, or by direct development by the local authority.'

But all this changed. The procedure for pressing forward with urban renewal was by declaring 'areas of comprehensive development' which proved to be exceedingly cumbersome and slow. It has often taken as long as five years for the whole procedure to reach the approval stage. Land prices are now far higher, but the methods of giving financial aid to local planning authorities under the 1947 Act were, one by one, torn away, making it almost impossible for smaller authorities and even difficult for larger ones to engage in comprehensive planning schemes. Local authorities had virtually no 'planning income' and were increasingly hard put to it to make capital available for redevelopment and to pay high interest rates. In particular, there was a very long initial period–before rents and rates begin to produce revenue–when no revenue was coming back for all the money outlaid. Yet local authorities still had to 'balance their books' annually during this period.

In such a situation the role of the local authority as prime mover in redevelopment gave place to the entry, on a grand scale, of developers who sought to redevelop the 'plum' sites because of the relatively higher financial returns which became possible. In less prosperous areas, or areas where there might be some element of doubt about good financial returns on money invested, it was much more difficult to induce commercial redevelopment; yet such unfavoured areas are just the areas, from the point of view of community interests, that most need such treatment.

For residential redevelopment the position has been even bleaker. It is often impossible to avoid a loss on residential development in inner city areas, even when extra subsidies for expensive sites or high buildings are taken into account. By

177

definition, speculative developers are not interested in making losses, so much of this development tends to be left to the local authorities to tackle; efforts to interest private 'enterprise' in urban renewal housing projects have, so far, met with little success.

These legislative, financial and administrative obstacles have been powerful brakes upon urban renewal and have been in contrast to the relatively favourable position of new towns where land could be bought at existing use values and finances are readily available by Government loans which also covered the initial unremunerative periods while planning and building were in progress. It has been possible for new town development corporations to balance their less remunerative development by retaining central area land in their possession and by themselves developing town centre sites, designing and building new shops and offices, leasing the property on rent and thus receiving back a measure of the increased values arising from their activities. In addition, new towns have few of the acute problems of renewing underground services, of demolition before building can commence, or of rehousing families and relocating industrial concerns displaced by development. With relatively unfavourable circumstances, it is not surprising that there is still so little to show in the way of urban renewal in Britain. Indeed it is surprising that so much has been achieved both in the examples discussed and elsewhere, despite all these formidable obstacles. A beginning has been made but it cannot be assessed as more than that. The scale of operations required needs now to be far bigger. The way ought now to be cleared for greater public and private endeavour, for urban renewal on a massive scale.

CHAPTER 7

Urban Centres

Until the end of the Second World War the town centres of Britain had changed little for 100 years. There had been some piecemeal renewal of individual sites but the general pattern was unaltered. The need and the opportunity for large-scale redevelopment of some centres was the result of bomb damage; towns and cities which had suffered severely were naturally the first to begin rebuilding. What was not anticipated was the surge of redevelopment in the centres of towns which had escaped damage. The origins of this movement, which assumed immense proportions by the mid-1960's, were the increase in consumer spending which led to demands for more shopping space, changes in retailing methods requiring different, and usually larger, types of shops, and increases in white-collar employment requiring much more office space.

Central area plans were drawn up for most of the blitzed cities though they did not in the main make adequate provision for increased use of cars; few of the other towns and cities had any plans so that redevelopment generally followed existing street patterns. The conflict between the desire, on the one hand, to have access for vehicles everywhere, and, on the other, to retain a civilized environment, is therefore most evident in urban centres. Functionally, economically and visually, the motor car now threatens all centres; physical danger, noise, fumes, irritating and costly delays are our daily experience in great cities and small towns alike; squares, streets and demolition sites are littered with parked cars. And, with an expectation of twice as many vehicles on the road within ten years, these conditions will be progressively aggravated.

At the same time, indeed, because of the increased use of cars, public transport is tending to run down. This is a complete reversal of the pattern of the past 100 years. The cities we know are the product of a highly organized community transport system. Without the underground and suburban railways and bus services, London could not possibly support its high density of office and commercial uses; our provincial cities, on a different scale, rely equally heavily on mass transport. This pattern is changing swiftly and radically. Figures given in *Economic Trends* issued by the Treasury in November, 1963, showed that in 1952 private transport carried one-third of all passenger movement in Great Britain; by 1958 private and public transport carried about equal numbers. Official estimates for 1965 indicated that passenger mileage by private car amounted to nearly 70 per cent of the total: the number of passenger journeys by public service road vehicles had fallen by 25 per cent since 1955. These trends are strong even in travel to and from major cities.

In Britain the size and nature of the problem is most evident in London. Yet it was not until 1962 that a comprehensive survey of traffic was begun. The first two volumes of a report on it have now been published.[1] The survey covered most of the area inside London's green belt, nearly 1,000 square miles, containing a population in 1961 of 8·8 million. Information was gathered by interviewing 50,000 householders, one-fifth of whom were chosen because they did not own a car, by roadside interviews and traffic counts, by collecting data from transport undertakings and by measuring journey times. All this information, processed by computer, provided an inventory of facts about travel patterns and volumes, the traffic capacities of existing roads and public transport services, and the generation of journeys of people and goods. These and many other figures are set out in detail in the first volume of the Report; some confirm observable tendencies, others are quite unexpected. For instance, there are 8·8 million vehicle journeys on each working day, almost exactly one per

[1] Freeman Fox & Partners, in collaboration with Engineering Service Corporation of Los Angeles and Wilbur Smith and Associates of New Haven, Connecticut, *London Traffic Survey*, Vol. I, 1964; Vol. II, 1966.

person in the survey area, and nearly all of these were move-ments within Greater London. The proportion of car owning households is directly related to total household income; 80 per cent of families with an income of £2,000 per annum or over have a car but only 26 per cent of those whose income is less than £1,000. Non-work journeys are very much higher for families that own cars than for those who do not; about two non-work journeys per day are generated for all London households as compared with three or four in American cities.

In Volume II of the Report forecasts are made of the pattern of travel demands in 1971 and 1981 and include a preliminary study of alternative road networks. The results suggest that work trips will alter little. Nearly all the increase will be in non-work trips; trips within the survey area for car-owning households will rise by 8·2 per cent but average trips for all households by nearly 40 per cent to 5·34 per day. Assuming a considerable investment in roads in the 1970's to provide London with a basic urban motorway network, saturation point in the year 2000 would be about 5·88 trips per day.

Whether people move about by public or private transport is, of course, greatly influenced by the type of transport available. The Report had to assume that known schemes for public trans-port would be put into effect; for roads two alternative schemes were assumed, one on the basis of increased expenditure in the 1970's to provide a motorway network (1981A), and another on the limited scale which would be the product of continuing expenditure at the present level (1981B). Assessments of the volume of traffic on the 1981A network were that all the roads forming part of it would carry 90,000 vehicles per day (a figure exceeded at present only by Park Lane) whilst considerable lengths of these roads would be required to carry 150,000–200,000 vehicles per day. Whether the required number of traffic lanes and junctions are a practicable engineering proposition was not examined. The calculations merely show the size of the probable demand: it was estimated that whilst rail traffic should remain fairly stable until 1981, bus trips will decline from 3·9 million to 2·7 million per day; private car trips will increase from

4·1 million to 9·0 million per day and goods vehicle trips by 73 per cent. The task of finding a socially and economically acceptable combination of different forms of transport was left for consideration in Volume III.

The size and complexity of this task is evident; that it is so vast and complex stems directly from past lack of policy on transport and traffic. The proposals of Abercrombie and Forshaw in 1953 appeared far sighted and even idealistic; in the light of twenty years experience it can be seen that they were too modest. And even those proposals were not carried out; in particular there has been little Government assistance towards improvements in the road system and a formidable backlog of work built up. Action has never measured up to need.

The central area of London is unique and cannot adequately be dealt with in this book. It is not only a question of size; the problem is a special one because of London's function as Britain's capital, as a focus of international commerce and finance, and as a shopping centre which, to a degree, serves the whole country. Added to this, the West End and the city are studded with historic buildings and sites not lightly to be interfered with, and redevelopment has been hampered at every turn by the complex system of local government and the need to obtain the approval of Government departments for many major projects. The administrative structure has recently been radically altered by amalgamation of the old metropolitan boroughs into larger units with wider powers; the Greater London Council has replaced the former London County Council. It is too early to judge whether these new arrangements will provide the co-ordinated planning which has so far been lacking.

The only major traffic improvement in the centre of London has been the Hyde Park Corner/Marble Arch scheme. By taking a slice off the park, it has been possible to make Park Lane a dual carriageway providing a relatively free flow of traffic. At the south end, where six major routes converge on Hyde Park Corner, two gyratory systems have been constructed; at the north end, the Marble Arch roundabout has been extended, and near it is an underground car park for 1,000 cars. Beneath the roads there are pedestrian subways approached by steps or ramps. Quite obviously, drastic measures had to be taken to relieve congestion at Hyde Park Corner and Marble Arch, but much has been lost in visual terms—for instance, the fine composition of Wellington Arch at the end of Constitution Hill has gone, and there is no longer the change of atmosphere given by driving through the park parallel with Park Lane. These losses—and the financial cost—are difficult to justify when the result is merely to transfer the congestion somewhere else.

More hopeful is the attitude of the Passenger Transport Planning Committee for London, which is considering further developments of the underground railway system in addition to the Victoria line now under construction. Although new underground lines cost £5 million per mile it has been calculated that they will be capable of carrying five or six times as many passengers as new urban motorways costing the same amount. The study of costing for the Victoria line did, in fact, for the first time, take account not only of estimated revenue from fares (which was insufficient to justify construction) but also the estimated savings to the community in surface-road congestion, accidents and other factors—the social benefits.

Building since the war has been immense, particularly in the form of multi-storey office blocks which add further to concentration in London by business firms. In spite of the valiant efforts of the Planning Department of the former London County Council, only here and there are there signs of overall planning. The largest and most impressive scheme was that for St. Paul's Cathedral precinct by Sir (now Lord) William Holford. His proposals, sensitively designed in great detail in three dimensions, were modified (not for the better) by the Ministry, and have been further compromised by the quality of the new buildings actually erected, to the extent that much of the original merit of the design has been whittled away.

Generally, since redevelopment began to gather momentum, inflated land prices and the dismantling of the financial basis of the 1947 Planning Act have stacked the cards in favour of speculators, whose concern has been to obtain the maximum return of

capital. This state of affairs has aroused growing public appre-
hension which, on one occasion at any rate, was effectively
expressed in the protests about the redevelopment of the Monico
site at Piccadilly Circus. The objections put forward by the Civic
Trust, and other organizations and individuals, led to the com-
missioning of Sir William Holford to report on the redevelopment of the Circus as a whole. With no clear traffic pattern
established for the West End and astronomical land prices to
contend with, a compromise, though a skilful one, was the only
possible outcome. In the first scheme which Sir William put
forward, traffic would no longer circulate in the Circus, but be
routed along the west and north sides of a piazza; pedestrian
circulation would be at three levels–underground in an enlarged
concourse, at ground level, and by raised walkways. Architecturally, the shape of the Circus would be tidied up into a
simple rectangle with an even roof line; higher buildings would
be set back with the exception of the 'vertical feature' marking the
'hub of the West End' and providing a possible site for some
illuminated signs. This report was not accepted by the Minister
of Housing and Local Government on the grounds that it
allowed only for an increase of traffic of 20 per cent whilst the
increase expected, if the approach roads were improved, would
be of the order of 50 per cent. This was not a valid criticism of
Holford's scheme, which had terms of reference limited to the
Circus and to an interim period only; it was not, and was never
intended to be, a long-term solution of the problems of the West
End as a whole.

In 1966 Lord Holford was appointed to make a further study,
this time with wider terms of reference. The area to be considered
was increased to a group of sixteen street blocks, mainly to the
north and east of the Circus. The brief asked for a study of the
redevelopment of 'Piccadilly Circus and its setting as a place of
public resort' to form a basis for detailed design and assessment
of economic implications. It was also laid down that a 50 per
cent increase in traffic volumes should be allowed for, and that
comprehensive redevelopment of adjoining Crown property,
including Regent Street, could be assumed.

Traffic requirements thus became the main determinant for the Mark III plans: as much unobstructed space as possible had to be provided for carriageways at ground level. The solution chosen (Fig. 45) proposes a major light-controlled crossing in the centre of the space between Swan and Edgar's and Great Windmill Street. From Regent Street and Glasshouse Street in the north-west five lanes of traffic would cross to the Haymarket and Coventry Street; from the south-west sector (Piccadilly and Lower Regent Street) traffic would divide and cross over into Shaftesbury Avenue or into the Haymarket or Coventry Street. On the west side there would be three lanes to serve traffic going north from Lower Regent Street, and, on the east, four lanes turning down Shaftesbury Avenue into the Haymarket, which would be one-way southwards. This scheme would allow for even more flexible arrangements if future traffic demands require them.

But though the brief required road traffic to be given first priority, the special function of Piccadilly Circus as a meeting place and focus of entertainment was seen as being of equal importance. A triangular pedestrian concourse underground was envisaged; above the traffic at ground level a deck was proposed, given over to pedestrian movement in every direction. Above this again on the London Pavilion site could be a conference hall and restaurant suite. The deck would have links to the adjoining Monico, Trocadero and Criterion sites which it was assumed would also be redeveloped in the fairly near future in the form of shops, offices, restaurants, flats and hotel. Escalators would be used to assist pedestrian flow between different levels.

The structural problems of carrying decks and walkways over the complex traffic intersections were squarely faced. Points of support, even with large spans, are very limited if traffic and pedestrian flows are to work smoothly and great engineering and architectural ingenuity would be needed to meet all requirements, especially bearing in mind the network of underground railway terminals and other services below ground.

The report presented by Lord Holford in 1966 emphasized that the illustrations are only diagrams; they did not show details of design as his earlier proposals did. Much would depend on

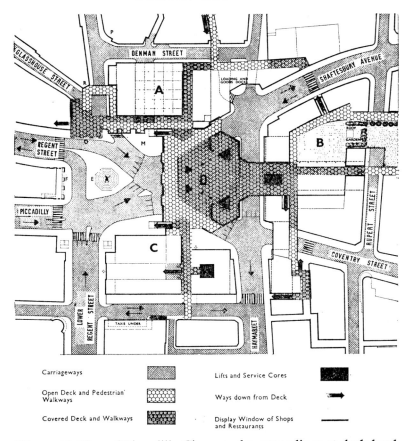

Figure 45. Plan of Piccadilly Circus and surroundings at deck level, showing pedestrian system.

A. Monico Building: B. Trocadero Block and Hotel: C. Criterion Block: D. New London Pavilion: E. Eros Statue and Fountain: F. Swan & Edgar: K. Monico Arcades: M. Flagstaff: O. County Fire Office: P. Piccadilly Theatre: Q. Lyric Theatre: R. Regent Palace Hotel.

how much of the scheme was carried out by the Greater London Council, how much by private developers and the degree of co-ordination between them. The proposals at any rate provide a coherent yet flexible theme for redevelopment by stages, and, given co-ordination between public and private effort, the bustle and glamour of Piccadilly Circus, as a focus for visitors from the suburbs, the provinces and overseas, could be retained in a completely new and better form, whilst the present architectural anarchy and traffic congestion could be eliminated.

TOWARDS A SOLUTION

Traffic problems cannot be resolved by piecemeal experiments. Comprehensive surveys like the London Traffic Survey are essential. They indicate the size and kind of problem which arises in each town; they show how people move at present and how new development and redevelopment will affect the present situation. But the collection and analysis of facts is not a substitute for thinking; we must also have a clearer idea of the urban pattern which is most suitable for present day life. In searching for it the main objective must constantly be kept in mind: it is not to provide better conditions for vehicles but to recivilize cities.

More and wider roads and more car parks will not solve the problem; traffic increases to fill the road space available. An analogy has been drawn up by Buchanan between the town and a complex building such as a hospital, the essence of which is its wards and operating theatres; corridors and lifts are a means of circulation–vitally important but secondary. The main elements should not serve also as circulation routes: it would be ludicrous for instance if food trolleys had to cross the operating theatre on their way to the wards. But circulation in town and city centres dominates the uses which it should serve. It has been forgotten that the centre is not primarily a place to which people and goods travel but in which people work, shop, meet their friends, and visit restaurants, theatres and concerts. The pedestrian is not just a nuisance and a hindrance to traffic; his or her desire to move about on business, look into shop windows, or just stand and stare, is the prime reason why the city centre exists at all. Even in London 35 per cent of the work journeys of central area residents are done on foot.[1] In great cities freedom for the pedestrian from injury, anxiety, noise and fumes can only be attained by placing footways above or below the distributory network of roads carrying delivery vans and essential private cars to shops and car parks. In smaller towns such complete separation may not always be needed; the closing of some streets to vehicles and provision of access to the backs of shops may be enough. But, everywhere, the long-standing concept of the heavily trafficked shopping street must be abandoned.

To disentangle the maze of streets, buildings and services is itself a formidable task but safety is not the only need of the pedestrian: the environment in which he moves must be worth looking at. In this sense, the car driver is not so important; he can only take in the broad picture. Pedestrians, moving more slowly, are more conscious of the relationship and detailed design of buildings and spaces, of 'character', of vistas and unexpected glimpses, of changes of level; it follows that only as a last resort should they be compelled to use subterranean tunnels which have few attractions and are no more than utilities for avoiding vehicles.

Completely new factors in urban design have been introduced by the growing popularity of high buildings resulting from the inflated price of land. But the disruptive effects of railway viaducts and cuttings are being reintroduced by urban motorways which are very much wider than railway tracks. Even with the retention of the principle of compactness these elements alone will change the texture of urban centres and the three-dimensional relationships of buildings and circulation space. The Roman agora, the medieval market place and the Renaissance square and street offer no precedents; new patterns will have to be worked out from first principles. Scientific considerations of daylighting buildings are only a preliminary step. The aesthetic aspects, which are really of sculpture on a gigantic scale, are a new and immense challenge.

The design and redevelopment of urban centres bristles with questions–functional, financial and aesthetic–to which there is

[1] *Traffic in Towns*, The Buchanan Report, 1964.

no easy, cheap or rapid answer. One of them–the conflict between traffic and environment–has at any rate been put into realistic focus by the Buchanan Report.[1]

The terms of reference of the Buchanan Working Group were 'to study the long term development of roads and traffic in urban areas and their influence on the urban environment.' The Report shows that the relation between roads and traffic and, even more, between traffic and other human activities does not regulate itself nor are these relations capable of being controlled by any existing regulative measures. Devices for doing so have therefore to be found, but before this can be done the problem needs to be stated in adequately comprehensive terms. The Report does this with admirable clarity:

(1) Traffic is made up of the individual journeys of people whether by rail, bus, car or on foot. Nearly all of these begin in one building and end in another–home to factory or office, home to school, warehouse to shop. The *density* of movements varies with the density of these buildings; the *complexity* of movement depends on the location of buildings relative to each other.

(2) All buildings require access of some kind and many of them need vehicular as well as pedestrian access. At the same time all buildings need, to a greater or lesser degree, freedom from noise, fumes, vibration and physical danger. A conflict arises in reconciling these requirements with those of vehicular access.

(3) Urban streets at present have to serve through traffic which has no reason to be there, local traffic which may have, and kerbside parking for people visiting shops or offices. They have also to provide for pedestrians whose movements will, at some points, cross traffic flows.

(4) The present conflicts between environment and traffic, and between vehicles and pedestrian movements, can be reduced but only within strict limits. Attempts to provide for free car access to all buildings show diminishing returns–a point is reached where roads and car parks occupy a

[1] Ibid.

disproportionate amount of land; insistence on very high environmental standards can reduce accessibility below a tolerable threshold.

(5) In weighing the claims of accessibility and good environment in any given area a further element has to be considered – cost. But a choice has to be made; if it is not, either accessibility or environment – and probably both – will fall below acceptable standards.

This problem can only be tackled within an overall strategy for urban design. The first step, quite clearly, is to take broad decisions about the ultimate size and rate of growth of cities and towns; it is impossible to design for needs which may change out of all recognition and cause sudden and incalculable demands. Once size and population are decided and accepted it becomes possible to make useful assessments of needs in terms of traffic, employment, shopping and so on. At this stage it is evidently necessary to study the function of each town and city in relation to its neighbours and the surrounding countryside, the relative 'magnetism' of shopping, business and administrative centres, the degree of competition between them.

Having done this we must go further and decide what kind of cities and towns we want and what form their centres should take. This is far from being a merely academic question – it is a very real and urgent one. As has been shown earlier, more and more people wish to travel into cities by car. Each month towns, large and small, become increasingly congested by private cars and it is more and more difficult to park. As the Buchanan Report pointed out, even in Coventry, where the problem has been taken seriously, only 7,500 parking spaces were planned for a population of 360,000, whilst the study of Leeds in the Report suggested a demand for 20,000 spaces for a population of 524,000.

The issue emerges, perhaps even more strikingly, in the study in the Buchanan Report of a section of the West End of London. Assessments were made on four different bases – complete, partial and minimum comprehensive redevelopment, and piecemeal redevelopment. With complete comprehensive redevelopment, the Buchanan team proposed a system of primary roads below ground, distributor roads at ground level and pedestrian circulation above. Twenty-three per cent of the site would be required for roads, loading and parking and even this would only allow parking for 20 per cent of the people working in the area. The scheme based on minimum redevelopment, with existing streets acting as local distributors, necessitates the use of all mews and access roads for the servicing of buildings, and movement would be difficult. Commercial movement at peak periods would have to be restricted and car parking space would be extremely limited.

The conclusion to be drawn for more general application is that, even in an area which can be redeveloped as a whole, unrestricted use of cars is impracticable if the present compact form of urban centre is to be retained. If complete demolition and rebuilding are impracticable, heavy restriction on the entry of vehicles is essential. And these are likely to be the most common conditions. There are relatively few urban areas which are completely outworn or contain no historic buildings which ought to be retained; even in these areas the difficulties of wholesale acquisition or the financing of total rebuilding will be very great. Space within city centres is expensive and the cost of building multi-storey car parks very heavy, so that, even if one takes account of the use of each car space by several vehicles for a short time each day, parking charges are apt to be high. The congestion resulting from cars leaving their parking places adds further to the peak-hour traffic flow on roads. And as more land is taken up in accommodating cars, walking distances to shops and offices become excessive.

It is true that much of the goods traffic into and around city centres is generated by uses which need not be near the centre and would, in fact, be better located elsewhere. With advantage to employers, workers and customers, many industries and some offices and businesses could move out altogether, eliminating the flow of traffic to them as well as reducing the concentrated peak loads of journeys to work. Even those uses which must remain in the centres of towns are often ill-arranged, and by their 'scatter' induce traffic which could be lessened by better location. But even radical replanning on this scale is not enough: a further

191

price will have to be paid for compact centres, in the form of permits to control entry of vehicles into certain defined zones, charges for entry and length of time spent in these zones, and provision of parking by the developers of all new buildings for essential traffic which they generate. Still more important–and probably most controversial–is the need to provide, for the majority of workers and shoppers, a community transport system, which is speedy, comfortable and very cheap.

LESSONS FROM THE U.S.A.

The first comprehensive attempt to study the economic aspects of various forms of urban transport was made recently in the U.S.A. by two economists and a transport consultant.[1] Their research covered thirty-nine of America's largest cities, excluding New York, and showed a very heavy degree of dispersal of industry, warehousing, showrooms, shopping and restaurants to peripheral urban rings. The availability of good and cheaper public transport had no effect on the continuous decline of central areas. Having found it possible to cross urban areas at 40 m.p.h. on new expressways at off-peak hours, people would like to do the same during rush hour periods. 'If they so insist, and if automobile ownership continues to rise rapidly, a need would appear to exist for many new expressways in urban areas. The only alternative would be a diversion of urban transport back to public transit.' On a purely economic basis the authors conclude that whilst the car is the cheapest or most convenient mode of travel where traffic volumes are low, city transit systems can be designed which are cheaper than travel by cars containing only one or two people; for commuter traffic from home to town centre, bus services are the most competitive; rail is only competitive where there are existing lines and in large cities with high densities. Meyer, Kain and Wohl generally favour increased construction of urban roads so long as they are effectively used: 'A lane of limited access highways reserved for buses can handle upwards of 25,000 or 30,000 seated passengers per hour at

[1] J. R. Meyer, J. F. Kain and H. Wohl, *Urban Transportation Study.*

35 m.p.h. to 40 m.p.h. while private automobiles even with full loads of five or six people would do well to handle about 10,000 people at the same speed in a single lane.' They show little enthusiasm for monorail but suggest the development of a vehicle which could operate both on rail and on highways; synchronized traffic lights, electronic control of buses and centralized fare collection are also advocated.

This study of conditions in the U.S.A. is instructive and valuable; in particular it stresses the old truth that the customer gets roughly what he pays for. The problems of the U.S.A. are, of course, not identical with our own; in Britain the car is, even now, not so much of a 'sacred cow', and urban decentralization has not yet occurred to anything like the same degree. Pictures of the car-dominated community in the U.S.A. have been forcibly presented by Edward Higbee[1] and Victor Gruen[2]; many wholesale and manufacturing establishments are being forced out of town centres by high rents and are moving to peripheral sites near main roads. Retail businesses which are not dependent on impulse buying or ready pedestrian access—cars, furniture, office equipment—are migrating because they need large floor areas and long-term parking. Suburban shopping centres grow at the expense of central areas. All these tendencies are linked to the use of the private car which, it is estimated, transports 74 per cent of the passengers in towns of 50,000 to 100,000 in the United States. Car parks are provided by public and private enterprise on a much greater scale than in this country and attempts have been made to relieve down-town congestion by locating them on the fringes and linking them with public transport to the centre.

Public transport, never so comprehensive as in Britain, has declined. The use of street cars in New York has fallen since 1950 by 78 per cent and the underground and elevated railways by 17 per cent. Left to itself public transport will die, but more and more proposals are made for improvements; San Francisco has been considering a scheme for rapid-transport trains; Cleveland has actually built a new rail system since the war. To speed

---

[1] Edward Higbee, *The Squeeze: Cities Without Space.*
[2] Victor Gruen, *The Hearts of Our Cities.*

pedestrian flows, 'moving pavements' have been suggested and are now being introduced in various parts of the world.

Two American planning schemes are particularly notable. At Fort Worth sweeping proposals have been made by Victor Gruen for reshaping the central area. He has suggested redevelopment in the form of pedestrian precincts with large parking garages, not more than three or four minutes walk from the centre, and accessible from perimeter expressways; delivery to shops would be by tunnels; slow-moving shuttle cars would provide public transport within the precincts. The legal and financial difficulties appear considerable; moreover, although 60,000 parking spaces are proposed, personal transport will not be enough; the scheme would only function if adequate public transport were provided – and only 17 per cent of those who now travel to Fort Worth use public transport. It seems unlikely in fact that the Fort Worth project will be carried out. There is a clear lesson here: it is uneconomic to provide large amounts of car parking and public transport as alternative modes of travel to a major shopping and business area; personal transport cannot carry enough shoppers. The major role must therefore be played by mass transport.

The second American example is Philadelphia where, since the Second World War, there has been intense activity and experiment in urban planning;[1] only the principal proposals can be dealt with here and in the broadest way. Philadelphia is the regional capital of an area whose present population of four million is expected to increase by 1980 to six million. It includes a quarter of the region's employment, all the main commercial and administrative facilities, its principal shopping centre and a focus of cultural activity of national importance. The basic plan of the city centre was laid down to a grid-iron pattern by William Penn in 1862. The schemes evolved over the past ten years envisage an interlocking system of three types of movement; vehicles on an expressway linked to parking terminals, railways and buses, and segregated pedestrian ways. The express loop road is intended to carry all through traffic and provide controlled access to the

[1] Eleanor Smith Morris, 'The Planning of Philadelphia', *Architectural Design*, August, 1962.

centre at points related to large parking garages. The underground railway is to be improved by lengthening strategic sections and putting some elevated sections underground. The suburban railway system, now consisting of two different lines, is to be connected underground and existing stations relocated. Parking for 17,000 cars is proposed, connected to the expressway and to the major passenger stations. Pedestrians will be able to walk from one side of the city to the other via a pedestrian mall and to reach passenger stations or business areas under cover in the existing or expanded underground concourses; a pedestrian upper deck promenade is proposed in the Market-East shopping complex (Fig. 46). In terms of environment the scheme is equally impressive and aims at a scale relative to movement on wheels or on foot, 'spaces which stir the senses', and continuity but variety

Figure 46. Market-East shopping centre, Philadelphia: sectional drawing. Three levels of retail shopping, the lowest linking city underground railway with commuter railway system.

in architectural form. The progress of this project should be worth close study in years to come as a pioneer example of forthright city renewal.

CITY CENTRES IN EUROPE

The famous Lijnbaan at Rotterdam (Plate 39) has been described and praised so often that it needs no more than a passing reference here; with its safe but lively atmosphere it has become the prototype for all new shopping precincts which do not rely on multi-level separation of vehicular and foot traffic.

In Stockholm, the new commercial centre was started in 1959. An area of 45 acres, mainly occupied by housing, has been cleared and is being completely redeveloped by the building of five eighteen-storey office blocks, many new shops and extensive underground car parking. Pedestrian shopping precincts are provided at ground level. At upper levels, roof gardens, cafés and a metropolitan park are linked by bridges. Road access to the centre is by two-level urban motorways which meet in an immense roundabout, a solution which may be effective for dealing with the traffic but seems to have serious aesthetic drawbacks in cutting up the city. But the importance of community transport has been recognized; the underground railway system has been enlarged and improved to carry 300,000 people per day; without this the traffic problem could not be solved (Fig. 47).

Shopping and business streets have been closed to traffic in many European towns including Stuttgart, Cologne, Essen, Amsterdam, Brussels and Paris. Probably the longest pedestrian street in Europe is the Strøget in Copenhagen. It is really five connected streets which wind for a distance of a mile through the medieval part of the city between its two main squares. In 1962 the municipal authorities and the police prohibited traffic for a temporary period of two and a half months. Traffic adapted itself to alternative parallel routes (which have now been made one way) with less difficulty than was anticipated so that the prohibition of traffic was extended to twelve months and later became permanent.

The whole length of Strøget is not completely free from traffic: there are main intersections at six points, three of which are controlled by traffic lights, and several other minor crossings. But the pedestrian is given priority throughout and the improvement in the environment is apparent; flowers, small trees, seats and an outdoor cafe have appeared in the street and parts of the carriageway have been paved over. The shops are serviced on each side by a network of narrow streets and most of the shop-

39. Lijnbaan, Rotterdam: prototype of shopping
precincts from which all vehicles are excluded.

40. Strøget, the main shopping street in Copenhagen: its character was completely changed by closing it to traffic.

41. Another view of Strøget in the evening near Christmas-time.

keepers seem to be satisfied with this. Car parking in the vicinity is limited; most of the shoppers come to Strøget by public transport on a route roughly parallel with it and a distance of from

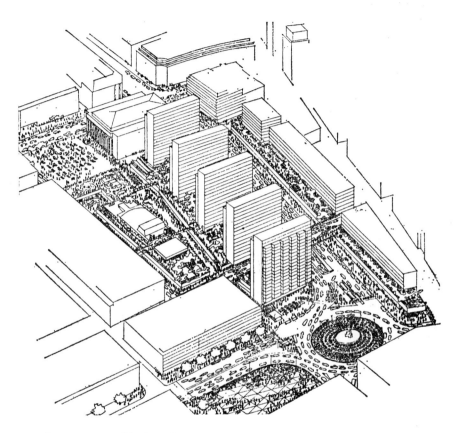

Figure 47. Stockholm: the central area showing the new office blocks and traffic circulation.

twenty to a hundred yards away. It is evident that exclusion of traffic has not led to loss of business; the majority of traders report increased sales; only one firm has moved away and did so for quite different reasons (Plates 40 and 41).

REGIONAL CITIES OF BRITAIN

Ambitious schemes for most of the regional capitals of Britain are now under way, adopting in varying degrees the principles of separating different types of traffic and pedestrians from vehicles. The ancient city of Newcastle, at the crossing of the River Tyne by the Great North Road, was among the first to take up the challenge of redevelopment on these lines. Until 1780, Newcastle was a medieval city focused on the riverside with steep narrow streets leading to the upper part of the town. The 'Great Scheme' for a new commercial centre, designed by Grainger and Dobson and carried out between 1835 and 1839, moved the centre of gravity of the city northwards. The coming of the railway created partial barriers between segments of the city; the Central station became a new point of attraction. But, apart from the building of the Tyne Bridge, the city changed its from little until recently, even though socially Newcastle strengthened its position as the regional centre of the North-East for shopping, business, education, entertainment and sport.

In 1960 the City Council set up a separate Planning Department under the leadership of Wilfred Burns. His first report, prepared with the active collaboration of other chief officers, dealt with the problems of central development. This stressed that any new development must aim at preserving and enhancing the character of the city. Since then work has gone steadily ahead. Traffic proposals based on large-scale surveys envisage four distinct but interlocking systems of movement and two types of parking accommodation. A motorway system round the central area offers the fastest and easiest route within it but will exclude traffic which need not be there; commuting traffic will enter the centre at a limited number of points, each having direct access to a capacious multi-storey car park. The central street system, consisting mainly of existing thoroughfares, will carry traffic which has business in the centre to service roads, bus stops and an inner cordon of short term parking places. The service road system will give access from the central streets to all buildings. The footway system—a wholly separate network of new lanes,

198

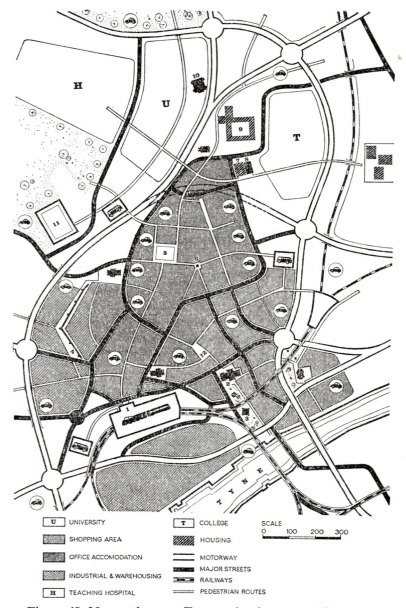

Figure 48. Newcastle upon Tyne: redevelopment policy map.
1. Railway Station: 2. St. Nicholas Church: 3. Castle Keep: 4. West Walls: 5. Eldon Square: 6. Grey's Monument: 7. Concert Hall: 8. Swimming Baths: 9. New Town Hall: 10. Hancock Museum: 11. Football Ground: 12. Market Square.

open spaces, arcades and existing streets closed to traffic–will provide for pedestrian access and circulation mainly at a newly created ground level. A 'moving pavement' may be installed on one of the busiest routes.

The need for adequate public transport is recognized and facilities are being provided for interchange between one form of transport and another. It would be quite impossible to provide access for the estimated four-fold increase by 1980 in the number of cars wishing to use the city streets; twice the present amount is planned for and a target of 13,000 parking spaces has been set.

The importance of Newcastle as a shopping magnet–its turnover is rather more than half of that of the whole of Tyneside–makes some expansion of the shopping area necessary; this is mainly to be linked to the present shopping streets and extended westwards. Provision is made for the retention of office uses in the present major office areas but some regrouping will be needed; the possibility of a small-scale removal of offices to outlying district centres has been considered.

The implementation of such a scheme obviously bristles with difficulties, not only the common ones of cost and relationship between public and private development, but the special problems of river crossings and fierce slopes which make Newcastle such an exciting place. But much has been done over the last six years and the pattern proposed in the 1960 report has been developed or revised in further studies and detailed designs; many existing buildings are incorporated in short-term or long-term proposals. The continuous process of planning and redevelopment is thus emphasized in these bold schemes for a city which has a wealth of historic interest and was unique in laying out a new business centre in the early nineteenth century. Close attention has been given to the importance of explaining to citizens what is being done and why it is being done: a series of booklets, admirably clear and informative, is available for everyone interested in redevelopment, and fuller information is readily given to those directly affected (Fig. 48 and Plates 42 and 43).

Manchester, as a planning authority, has problems which are

peculiar to itself. The population within its administrative boundaries is 700,000. The city is also the regional centre for two million people, many of whom live in areas which are contiguous with it; indeed the boundary of the county borough of Salford runs close to the centre of the city.

The plan for Manchester[1] published in 1945 was one of the most comprehensive of the reports which appeared in the years following the Second World War; it formed the basis of the development plan submitted to the Minister in 1951. It was not formally approved until 1961 by which time the pressure for rebuilding had grown and traffic conditions had worsened; it was apparent that the proposals for the city centre needed reappraisal. Surveys carried out in 1962 by the combined efforts of the City Engineer's Department, the City Transport Department and British Railways showed that about 35,000 people came into Manchester by train on each working day, 100,000 by bus, and about 15,000 by private car, assuming an average of one and one-third people per car. If all commuters had travelled by car there would have been an increase of twelve and a half times in traffic on the roads and a fantastic demand on space for parking if the normal 250 sq. ft. per vehicle is allowed. Studies showed that the greatest number of car commuters that could be accommodated was 20 per cent; to do even this posed difficult design and environmental problems, which, when studied in detail, involved material changes in the character and alignment of the primary road network proposed in 1951. Later reassessments of the traffic generated by existing and proposed land uses in the city centre have confirmed that it would be virtually impossible to provide for a higher proportion of commuters to come in by private car.

The basic assumption is that the majority of commuters can and must be carried by public transport and that shoppers and those wishing to transact business in the centre have prior claims on car parking space. The needs of these occasional or short-term visitors can be met if parks for them are not opened until 10 a.m. and are closed during the evening peak period between

---

[1] Rowland Nicholas, *The City of Manchester Plan, 1945.*

4 p.m. and 6 p.m. Ultimately 40,000 parking spaces can be provided in large units strategically placed with regard to the main road pattern; these will be under the unified control of the City Council and leased by them from private developers where necessary.

Great advances have been made since the formation of a separate Planning Department in 1963 under John Millar. From the first the new department was under heavy pressure in dealing with private redevelopment proposals and the Council's desire to accelerate slum clearance and redevelopment of the centre. The 1945 scheme for the central area reflected the over-rigid and formal ideas of that period and did not provide for the separation of pedestrian and vehicular movement; it had to be replaced by sketch master plans based on a pattern of linked pedestrian squares, in the main at ground level, using the natural falls of the site to allow for parking and servicing below.

The demands of the University, the Institute of Science and Technology and the United Manchester Hospitals, for land near the centre on which to expand facilities for higher education, also raised formidable problems. These bodies were brought together by the City Council in 1964 on a joint committee and a comprehensive plan for an 'education precinct' has been prepared by their consultants, Hugh Wilson and Lewis Womersley, which will provide ultimately for 28,000 students, of whom nearly 10,000 will be resident on the site.

The pragmatic approach which was necessitated by the pressure of events has been superseded by broader strategic methods. A rapid transit study is now under way; in parallel with this the City Planning, Highways and Transport Departments, together with the Ministry of Housing, are reviewing the entire field of transportation. It is accepted that a better mass transit system cannot 'pay' in the usual sense and will have to be subsidized to a greater or lesser degree: the aim is to find the most effective combination of public and private transport and the part that could be played by bus, railway or monorail. Road proposals are also being reviewed. The ring road scheme has been modified to include a western tangential road, and Mancunian Way to the south of the centre is to be supplemented by another distributor road closer in. Other proposals will require further investigation when the transportation study is complete.

Ambitious comprehensive development schemes have been prepared to replace the outworn and inadequate building fabric and road framework in the centre with a pattern of covered, air

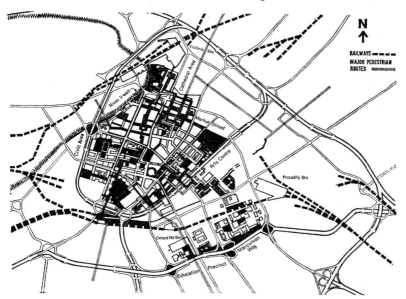

Figure 49. Manchester: central area redevelopment proposals: the areas shown in solid black are pedestrian and open spaces.

conditioned malls and pedestrian ways, the most important of which would be Market Street. But 'comprehensive development area' procedure under the Act is cumbersome, trained staff is always limited, and the Council, unlike those of some other major cities, owns little of the land in the centre. All these factors have hampered progress but a good deal has been achieved, including a substantial amount by agreement with private developers.

A new arts and entertainment centre form an important part of the civic area proposals; open spaces are planned along the rivers and canals and on derelict land. The Central station is to be closed; it is proposed that the great train hall should be retained as an exhibition centre and form the nucleus for re-development over a larger area with something of the character of the Tivoli gardens in Copenhagen. The wholesale meat and vegetable markets are to be moved out of the centre so that land will be freed for fairly high density housing accommodation. In addition to this vast central area programme, redevelopment of residential districts is going ahead as described in an earlier chapter. There is no doubt that Manchester today is setting about its problems with renewed energy, imagination and idealism; progress is greatly facilitated by the close collaboration of all the city departments concerned.

Liverpool's City Centre Plan[1] is a really full-blooded attempt by a big city, the centre of a large city-region, consciously to apply the principles enunciated in the Buchanan Report. Within a tightly packed area of one and a half square miles, the city centre serves as the main shopping, entertainment, commercial and business centre of the Merseyside region containing two million people; Liverpool is also Britain's second most important port. The plan aims at the continuance of these activities, which implies the renewal of about two-thirds of the city over a relatively short time and therefore a vast improvement of the quality of environment.

The solution of transportation problems is a vital part of the plan. Liverpool at present depends heavily on public transport; at peak hours 60 per cent of the people hurrying to and from work in the central area come by bus, 25 per cent by train and 15 per cent by car. The key decision made by the City Council was that public transport must continue to play a major part. For the commuter there is to be heavy reliance on an extended rail system but with express bus services on the primary road system. The possibility of using both private cars and public transport for different parts of journeys is acknowledged and moderate increase is envisaged in the proportion of journeys made entirely by car, but public transport will have the first claim

[1] *Liverpool–City Centre Plan*, Liverpool Corporation, 1966.

42. Newcastle upon Tyne: model of central area redevelopment viewed from west: St. Thomas Church on left, Eldon Square and Markets on right.

43. Newcastle upon Tyne: model of east central redevelopment viewed from south showing urban motorways: Civic Centre and St. Thomas Church top centre.

44. Liverpool: general view of the model of the redevelopment of the centre showing the intricate system of traffic routes: Mersey Tunnel entrance (*left centre*): River Mersey (*top right*).

on the city centre streets, and this will be upheld by control of car parking.

The proposed rail system is based on two major proposals. The first of these is for an outer rail loop encircling the suburbs and crossing the centre of the city underground, linking two of the main line stations. It is to be an electrified system and use existing tracks for most of the way. The second proposal is to link all four terminal stations together by an underground rail loop. This would substantially increase the capacity of the Mersey Railway, which at present runs under the Mersey to Birkenhead and the Wirral. It would also make for easy interchange between main line trains and suburban and regional lines through the stations linking the outer rail loop and the inner city terminal loop. It is estimated that the cost of these two proposals would probably be only about £10 million and for this relatively small sum the peak flow traffic problems of Liverpool could largely be dealt with (Fig. 50 (b)).

The present road pattern is a familiar one. There are sixteen distinct radial roads which not only connect the centre with the suburbs, but serve also as local roads and, towards the centre, as shopping streets as well. The Mersey Tunnel is a well-known bottleneck, causing city centre congestion of traffic within a quarter of a mile of the tunnel mouth; through traffic and city centre traffic persistently jam each other. There is no combined terminal for bus interchange with efficient waiting room facilities, etc. In all the main shopping streets there is serious conflict between pedestrians and vehicles.

The main proposal for the solution of the road traffic problems is for an inner motorway bounding the central area and carrying traffic between different parts of the city, freeing the central area for vehicles on local journeys. This inner motorway will be elevated for most of its length and connected to six main primary routes. There is to be a second Mersey road tunnel which will surface outside the inner motorway and connect directly to it.

The western sections of the inner motorway will have three lanes in each direction and will, for most of their length, take the form of a double-decked elevated road. This will continue along

(a) Proposed car parking—dark tone indicates major car parks accessible from the motorway. Lighter tone indicates areas within which car parking is accessible from the ground level street system. 1. Old Hall Street/Moorfields area; 2. Civic and social centre area; 3. Vauxhall area; 4. South City area; 5. Rodney Street area; 6. London Road area.

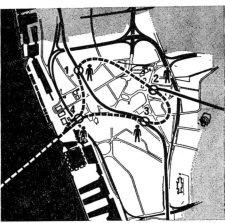

(b) Proposed rail system—The Mersey Railway and terminal loop is shown by a broken line. The outer rail loop and British Railways' connection to Lime Street are shown solid. 1. Exchange Station (low level); 2. Lime Street (ground and low level); 3. Central Station (low level); 4. James Street (low level).

(c) Proposed pedestrian network—Precincts and pedestrian routes are shown black. Major 'soft' and public open spaces are shown in tone.

Figure 50. Liverpool City Centre Plan.

the western length of the motorway, parallel to the Mersey, and then swing eastward, still double-decked for a short length and join in with the South Lancashire Motorway. Traffic will flow unimpeded along the western and southern sides of the loop, but on the eastern and northern sides it is proposed to install traffic signals to deal with the many connections that will be needed. Thus, in the space of a few thousand yards, there must be slip-roads to and from two major car parks and connections with four main primary routes. The use of traffic signals has been accepted in order to permit the scale of the intersections to be reduced, but it will have the effect of slowing driving speeds below those on the other side of the loop.

The plan emphasizes the importance of integrating the design of the motorway and of the urban fabric on either side of it, seeking to avoid mistakes elsewhere with urban motorways driven ruthlessly through existing urban areas with disastrous visual and social consequences. The Liverpool inner motorway is aiming at an entirely new kind of townscape in which the architecture of the motorway and the buildings on either side of it are conceived as a total environment. In some cases it is suggested that it may be possible to integrate buildings with motorway structure by building warehouses under it and car parks and other buildings over it. But all this can only be achieved by comprehensive re-development in depth on either side of the motorway. For these reasons parliamentary powers have been sought to acquire wide areas on both sides of the road as well as land for the motorway itself to be built on.

The plan adopts a construction target of 21,000 new public off-street car spaces by 1981. The total number of spaces likely to be available in 1981 would then be about 27,000. It has been envisaged that about 12,000 of the total number of spaces for cars will be used by commuters, 7,000 by shoppers and 8,000 for business trips and for the collection and delivery of goods by car. The total of 27,000 car spaces has been dictated by the expected capacity of the inner motorway and its associated radials and by the main street network, after allowing for through traffic and buses (Fig. 50 (a)). But it has been estimated that by 1981 the demand for parking in central Liverpool will amount to 64,000 spaces. To reduce this demand to match the 27,000 places that are likely to be available, the Council has adopted a pricing policy based on variable hourly charges. The adoption of this regulator means, of course, that the plan must provide for parking spaces not only at the most accessible places, but near to those parts of the city where drivers will be most willing to pay to park. Surveys of parking behaviour have suggested that this will favour shopping areas, since visitors to the city centre shops intending to stay only for an hour or so occasionally will be prepared to pay a higher hourly charge than commuters faced with paying to stay all day throughout the working week. So more parking spaces are to be built within walking distance of the shops than the offices, although it will be essential to provide some spaces near offices for visitors. Twenty thousand of the parking spaces will be provided in four major structures, each containing 5,000 spaces and directly accessible from the inner motorway so that it will be possible to drive straight into them.

Certain conclusions have been arrived at in considering parking policy in the central area about how far people can be expected to walk. The Report concluded that an acceptable walking distance from car to workplace for a commuter was between a quarter and a third of a mile. Shoppers, on the other hand, would not be expected to walk more than 300 yards from car park to shopping centre. These factors have been borne in mind in planning the pedestrian system. The basic aims of the plan, in providing for the pedestrian, are summed up in the statement: 'Although virtually all trips in buses and cars start and end on foot, it does not follow that footpaths have to be next to roads. Indeed, in a modern city, they should be as distinct and separate as the walkways and canals of Venice.' So the existing streets and pavements of Liverpool have had to be completely reorganized. Fortunately, the scale of redevelopment is so great that this can become a practical reality in the next fifteen years. As development is completed, additional lengths of high or ground level walks will be added to a growing network. Where rebuilding is not to be comprehensive, the old pattern of streets will remain

45. Liverpool: a detail of the model of the central area: St. George's Hall, the Art Gallery and the main pedestrian piazza.

46. Redevelopment Scheme for the Clock Tower and Haymarket Area, Leicester, photograph of model. *Left*, pedestrian piazza with tower of prestige offices and stepped walkways around. *Right*, the triangular block containing a comprehensive shopping and entertainment centre, to be built in the next four years. *Top left*, cluster of towers comprising luxury flats.

47. The Lower Precinct at Coventry looking towards the tower of the old Cathedral: two-level shopping.

and it will be necessary to restrict some of them to pedestrians and some to vehicles. The high walks of the rebuilt areas and the pedestrian streets of the existing parts of the city will, taken together, eventually provide a complete network of pedestrian walkways throughout the central area. Where it is necessary to climb noticeably to get from ground to upper level walks, speed ramps, escalators and lifts are proposed (Fig. 50 (c)).

The publication of the plan in book form is a major achievement in public relations; the complex process of planning is made clear and indicates that all aspects have been considered. The only big question mark is not over Liverpool but over the nation – will the necessary minimal government financial aid be given to support the proud efforts of the city fathers and their highly skilled planning team? The restraints of cost are recognized and compromises are accepted, but there comes a point beyond which compromise is impossible without vitiating the objectives of the plan altogether.

The establishment of a new City Planning Department in 1962 provided stimulus for a fresh look at the planning policies of Leicester. After one and a half years work a traffic plan emerged; it would be more accurate to describe it as a long-term urban policy plan with the stress on traffic.[1] The plan deals with the whole urbanized area of Leicester, reaching beyond its administrative boundaries. It is intended to cover a period of thirty years during which time the population of the city would rise from 270,400 to 323,000, and of the whole urbanized area of Greater Leicester from 453,000 to 640,000. The plan is based on a comprehensive public and private traffic survey, linked with studies of land use and the social and economic structure of the city. Instead of counting cars, people were interviewed in there homes, and for this purpose about 100 women interviewers were employed: an electronic computer was used in processing the material. The survey, which took about a year to complete, led to the conclusions that within thirty years the present number of cars would increase four-fold, almost to a saturation level of 1·2 cars per family.

W. K. Smigielski, *The Leicester Traffic Plan*, 1965.

The chief source of conflict in the present city is the lack of balance between the land uses generating traffic and the available traffic accommodation, roads and parking spaces. Unless speedy and drastic action is taken it is apparent that within thirty years the growing lack of balance will cause complete breakdown. A study was first made of the effects of allowing uninhibited use of the private car in all parts of Leicester including the centre. This produced a pattern of elevated roads and huge car parks surrounding a small shopping and business centre; the cost would have been £400 million, without including the cost of local distributor roads. The urban environment would have been destroyed: another solution had to be found.

The alternative proposed was an integrated transport system depending on the balanced interplay of a number of elements. Interchange car parks, attached to radial roads and linked to suburban shopping and entertainment centres, would have the function of intercepting private car drivers and taking them into the central area by modernized public transport. This interchange system would reduce the density of traffic on the radial roads and inner motorway; the inner motorway, only five miles long but of high capacity, would replace the three ring roads with a much greater mileage proposed in the earlier plans. Public transport would comprise a monorail, serving Leicester as the Underground does London, and three types of buses. In the central area, circulation would be facilitated by a special type of bus for standing passengers only, and moving pavements would be installed on the busiest pedestrian routes in the form of elevated bridges and walkways; a new type of small electrically-driven taxi has also been considered. The entire scheme is based on integrated transport as a public service and not a commercial enterprise. Vehicle penetration to the centre would be limited to essential business and commercial traffic together with 22,500 parking spaces, mainly for short term use.

This is a bold and ambitious project for the gradual redevelopment of an historic and prosperous city, without revolutionary transformation but aiming for the maximum level of circulation at the minimum cost, and far more attractive conditions for living, working and shopping. To what extent it will be put into effect remains to be seen. At any rate the survey techniques adopted are worth close study since they could be applied and developed by other city planning authorities.

Detailed studies in three dimensions have been prepared for the redevelopment of various parts of Leicester's central core. One of these is for the Clock Tower area, where traffic congestion is at present particularly acute; indeed, in spite of regulation by police throughout the working day there is continuous confusion and chaos. Comprehensive redevelopment is essential not only because of the traffic problem but because site values are high; the surrounding buildings have little architectural merit and some are obsolete. The traffic plan proposed would release this area from through traffic; it has been redesigned as a pedestrian precinct with stepped walkways around it. Shops on ground and first floors with an entertainment centre above are the main uses, thus reflecting the fact that the Clock Tower is the real commercial centre and meeting place of Leicester (Plate 46). The triangular area to the east comprises shopping arcades with a theatre, ice rink, cinema, dance hall, hotel, restaurant and roof gardens above. Underground service roads will provide for delivery to shops; a car park for 500–600 cars will be approached by ramps from the rear. This will be the first phase of redevelopment and is planned to be completed in four years.

The central piazza is conceived as a formal urban space; the Victorian Clock Tower is to be preserved as a matter of sentiment rather than for its intrinsic merit. The tall tower on the west would contain offices and the smaller towers to the northwest luxury flats. The piazza would be linked by pedestrian ways to the Market Square and other main shopping streets.

COVENTRY

The problems of towns which are not in the fullest sense regional centres are different not only in degree but in kind. Among themselves they vary immensely in size and in their influence on the area which adjoins them. Most are within the magnetic field of a regional centre such as Birmingham, Glasgow, Manchester or Bristol; at the same time they have their own fields of influence in the more immediate vicinity.

At the upper end of the scale is Coventry. Until the Second World War it was basically a town with a medieval street pattern which had missed the main impact of the Industrial Revolution but had grown rapidly as a result of the introduction of modern industry, particularly associated with car manufacture. Just before the war, a plan for redevelopment of the centre was prepared by the City Architect, (Sir) Donald Gibson, one of its main features being a pedestrian shopping precinct, the first in Britain.

The existence of this plan made it possible to persuade the Government to allow redevelopment to get under way on a comprehensive basis after the tragic destruction of much of the city centre by concentrated air attack. In its response to the challenge presented – the continuous drive and enthusiasm of its Council and their professional advisers – Coventry took pre-eminence among the cities of Britain. Traffic, housing and central area redevelopment have all been tackled with vigour; what is just as significant is that the town planning, architectural, engineering, financial and other aspects have been closely co-ordinated. This is planning in its proper sense; the result has been more effective action and more rapid progress than in any other British city of comparable size and in many which are much larger.

Coventry's problems have changed materially since the preparation of the 1938 plan and its population has grown to 337,000. The plan has been, and is still being, modified to meet these new conditions. This is relatively easy because the original concept was simple and is therefore capable of continuous improvement and adaptation to an extent which would not have been possible if a more complex and less flexible basis had been established.

Factories are well dispersed over the city so that industrial traffic and journey to work in the centre are less of a problem than in some other cities; there is a relatively small proportion of office employment in the centre which also lightens the load of peak hour travel. Nevertheless there is a considerable journey to work movement between various parts of the city. The latest road proposals reject the earlier concept of a series of ring roads. There is still to be a ring road round the city centre but the primary road network now includes an inverted 'Y' shaped urban motor road system tangential to the inner ring and relieving it of traffic which has no need to be in the centre. The eastern leg of the 'Y' will provide a link with the M1 to London, the M6 to Birmingham and the north-west of England and the A46 to Warwick, Leamington and Stratford-on-Avon; much of this road will be routed along a disused railway line.

No firm decision has been made as to the future mode of public transport; a transportation study is being carried out taking account of the needs of the ring of towns around Coventry as well as those of the city itself. Pending the results of this survey it is thought that mass transport will continue to rely on bus services, perhaps running in separate lanes near the kerb on the busiest parts of their routes. Other possibilities, such as a monorail system, are not being overlooked but Coventry has benefited so much from having a flexible plan that there is a very understandable bias in favour of a mass transport system which also will be adaptable to changing needs.

The central area is, of course, world famous as the pioneer example of a large modern pedestrian precinct in Britain. At an intermediate stage of its development, pressure from shopkeepers led to a proposal for a traffic road cutting across the precinct halfway down it. After Arthur Ling took over as City Architect and Planning Officer in 1955, the drawbacks of this compromise were recognized and the plan was changed back to the original basis – complete exclusion of traffic from the shopping area. The scars of this controversy remain, in particular the north–south precinct is excessively wide, but this defect has now largely been overcome by placing kiosks and other low structures and incidental features between the shop façades.

Eventually pedestrians will have priority over the whole area within the inner ring road; it will then be possible to traverse on foot the whole city centre, from the Cathedral to the market. Yet walking distances need not be excessive – the ring road bus route

213

will never be far away, moving pavements and escalators can be installed if needed, cars will be near at hand in multi-storey parks or on roof tops. Though surrounded by traffic the central area

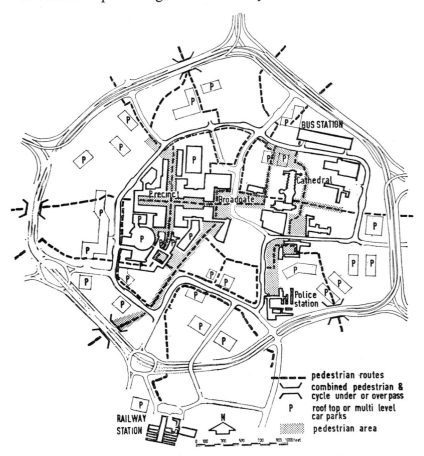

Figure 51. Coventry: scheme for city centre redevelopment: traffic circulation pattern.

will still be a place where the pedestrian is free to hurry or loiter as he will. Flats are now being built above the shops–a return to older urban customs which will add much to the liveliness of the centre in the evenings when the shops have closed.

214

These flats are in the form of towers which give added and necessary scale to the heart of the city and are part of a pattern of towers complimentary to the traditional 'Three Spires'.

The centre of Coventry is a striking example of the process of continuous change and renewal. It is not complete; in a sense it never will be, since the process of change will go on. It is fascinating to observe over the years a growing confidence and subtlety in working out details. Already it is a major achievement in terms of architecture–and in its atmosphere, which combines safety with bustle and vitality. It is paradoxical that the hub of the British car industry will probably be the first to give pedestrians freedom of movement in its centre (Plates 47, 57 and 58, and Fig. 51).

CUMBERNAULD

New towns have provided fruitful ground for experiments to solve traffic difficulties in town centres. In the earlier ones, such as Stevenage and Harlow, bitter opposition to the exclusion of traffic from town centres was only slowly and painfully overcome. The development of the shopping centre at Stevenage demonstrated that the theory of vehicular/pedestrian separation is a workable proposition from the shopkeeper's angle. The long-term profitability of the centres of new towns has been proved by experience; there is now a firmer basis for further experiment.

This is being made at Cumbernauld, in Central Scotland, where the aim has always been a compact town with complete pedestrian/vehicular segregation. Compactness in the centre was desirable because so much land was required to deal with vehicular circulation; it was also regarded as a virtue in providing protection from the weather for the new centre in its exposed position on the ridge which dominates the area. For the centre of Cumbernauld, multi-level separation has been adopted; the section (Fig. 52) is the key to the scheme. The lowest level is to be bisected by a dual carriageway; here will be unloading bays for shops, a bus station and parking for 3,000 cars. The upper levels,

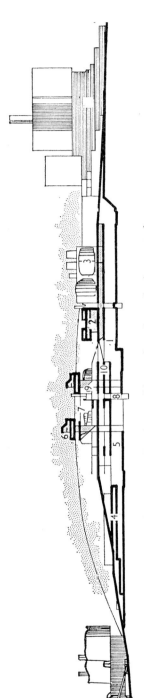

Figure 52a. Cumbernauld: section through the central area.

Key: 1, three-lane carriageways; 2, hotel; 3, government buildings; 4, office decks; 5, parking; 6, penthouses; 7, crèche; 8, service cove; 9, medical building; 10, central decks.

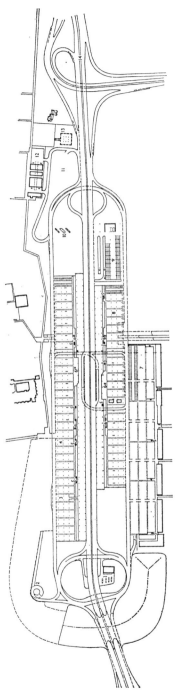

Figure 52b. Cumbernauld: plan of the central area at road level.

approached by escalators, lifts, stairs and ramps, are laid out as pedestrian squares and terraces, and above the shops and offices will be maisonettes, entertainment buildings, government and local authority offices, churches and educational buildings. The first stage of the scheme was started in the autumn of 1962 and is now virtually complete. For a town of 70,000 people, this is a more radical solution of the traffic and environmental problem than has so far been attempted; a few years ago it would have been thought impracticable. The progress of the experiment will be of immense interest and value.

THE SMALLER TOWNS

Central area problems are not confined to great cities but are also encountered in smaller towns. Many of these have stagnated since the hey-day of the railway age, nearly all have an intricate mixture of uses at their centre, housed in inconvenient or decaying buildings. Long-distance traffic and cross-town traffic struggles along narrow streets which are also expected to provide kerbside parking. And at night the life of the centre dies because few people now 'live over the shop'. The old country and market towns are rich in charm and architectural interest; nearly all towns in Britain, even those which are not beautiful, have robust individual character; all have established patterns which cannot be ignored or lightly interfered with.

The need for renewal and reshaping is just as great in these towns as in the larger ones; the problems are less complex but the means of meeting them are smaller in total scope and size and they are also, in some ways, of a different kind. The smaller towns serve, to a greater or lesser extent, the district around them, but are usually within the magnetic field of a nearby city. Their municipal authorities are seldom wealthy; they usually do not own much property in the centre; professional staff for the preparation of long-term schemes is often not available and planning powers are to a greater or lesser degree in the hands of the county council. Public transport is also usually outside the control of the smaller local authority.

217

Before these smaller towns can have efficient and pleasant centres many questions have to be asked. Should the core expand or will the weeding out of unsuitable uses, such as industry, provide enough extra space? Can people be brought back to live in the centre? Are the shopping areas, the town hall, the bus station well related and conveniently located? Which buildings and what elements of character should be conserved if humanly possible? And, running as a connecting thread through all these questions, how does the circulation pattern work now, how can through traffic be diverted, whilst allowing for delivery to shops and private parking? It has been said that something like 600 small towns are engaged in replanning parts of their central areas or considering projects for pedestrian shopping precincts from private developers. One cannot help wondering in how many of these have the questions outlined above been answered or even studied. The temptation to obtain rapid and apparently painless increase in rateable value is very great; piecemeal redevelopment of parts of a town is seized upon as a solution to some obvious current difficulty. But a few shortsighted decisions, without an overall plan, can make a workable town centre virtually impossible to achieve in the future. It is, in fact, worth while making a brief survey of the more comprehensive replanning schemes for smaller towns as an indication of what is being done.

Many town centres take the form of a linear main shopping street with side streets branching off at right angles. An example is Sutton, Surrey, for which redevelopment proposals have been worked out by the County Planning Department and the Borough Engineer's Department. The population of the present borough is of the order of 80,000 and is expected to increase by 13,000 by 1981. The High Street is not only the main shopping street but a major through route for north–south traffic: it is crossed by an important east–west route as well as by a number of minor streets. The relatively low proportion of outworn buildings in the central area made complete redevelopment impracticable. Instead it was decided to retain the linear character of the High Street and to divert the main traffic routes on to parallel channels to the east and west connected by gyratory systems at the north and south ends of the central area. Shops, offices and flats are to be built on the large areas of land enclosed by the gyratory systems, beneath

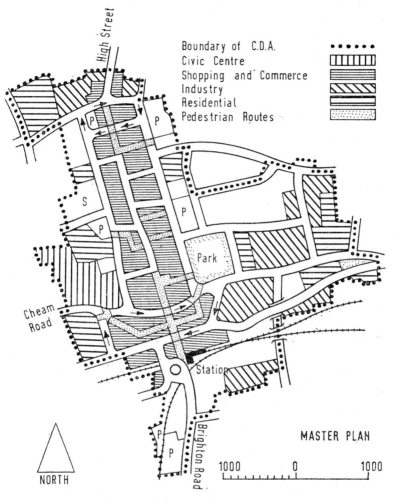

Figure 53. Sutton: the central area.

which are to be pedestrian subways providing links with the rest of the town centre. The whole of this traffic circulation pattern has been based on a very thorough survey and analysis of traffic

219

movement at different times of the day and on different days (Fig. 53).

The High Street will therefore become mainly a pedestrian street; only a portion of it will remain open to local traffic. There will then be clear views down the High Street to the central London skyline beyond. Groups of high buildings are proposed at strategic points at either end of the High Street and around the large open space to the east. Moderate shopping growth and considerable office growth is allowed for; good communications make Sutton attractive for office decentralization from London.

Very close attention has been given to public relations. During the two years spent in preparing the plan all the public bodies and private persons affected have been consulted and an exhibition has been held to enable everyone interested to study the reasoning leading to the proposals and their impact on the town.

Tamworth, on the boundaries of Staffordshire and Warwickshire, is an important junction on the main railway lines from London to the north of England, and from east to west, and a vigorous shopping and business centre. Its population is to be increased from 33,000 to 78,000 by the rehousing of people from Birmingham; it serves adjoining agricultural and mining districts with a further population of nearly 10,000 which is also likely to increase. The centre is hemmed in between the River Anker, the railway and the Birmingham–Nottingham trunk road; other main routes run through narrow shopping streets congested by kerbside parking. There are many fine old buildings but many are in an advanced state of decay.

In 1960, when the target population for Tamworth and its surrounding district was set at 50,000, the Borough Council commissioned a redevelopment plan for the town centre. This scheme, prepared by John Tetlow and Partners, is illustrated in Fig. 54. The trunk road is to be relocated clear of the town centre but the roads to and around the centre would still be inadequate for the growing volume of traffic, particularly in a south-easterly direction. Long-standing proposals for relief roads to the east and south were incorporated in the plan to form a ring round the central area. Within this ring a new bus station

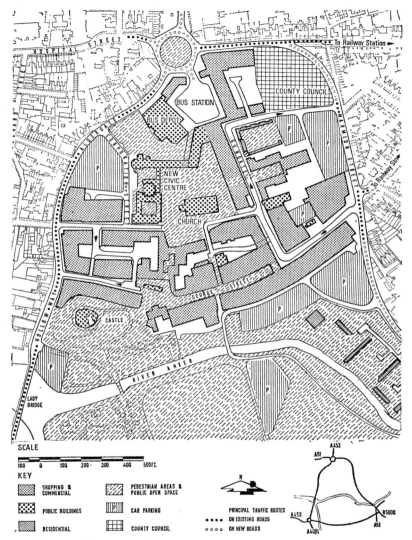

Figure 54. Tamworth: central area redevelopment scheme:
main road system inset (Plate 50).

and car parks were proposed; some of the car parks could become multi-storey if required, none of them more than 200 yards from shops. For the shopping streets themselves, the consultants originally considered a two-level solution, which is indeed suggested by contours. This would have entailed extensive acquisition of property and comprehensive reconstruction which proved to be beyond local resources. It was abandoned in favour of a scheme closing one of the main streets to traffic, providing access to the rear of all shops and a large measure of pedestrian separation, which it may later be possible to improve further. Attention was given to providing good settings for the notable buildings of the town, retaining its compact character and yet ensuring visual links with the open land to the south. The scheme was publicized in the press as soon as approved by the Council, and by free leaflets and an exhibition which included a model (Plate 50 and Fig. 54).

Recent decisions that the population of the borough shall increase to 78,000 and rethinking on regional traffic patterns make it necessary to examine Tamworth's problems anew in 1968. Nevertheless the 1960 plan has served its purpose in guiding redevelopment along lines which offer no obstacle to the modifications which will be necessary.

Newark, at one of the main crossings of the Trent, is another fine old town where the centre is threatened by disintegration under the pressure of traffic. Although a bypass of the A1 trunk road has provided some relief, traffic on the A46 still has to struggle through the centre, and congestion is aggravated, particularly in the summer months, by traffic on the A17 bound for the east coast. Local circulation and distribution is seriously impaired; parking for private cars and service access for shops is inadequate; there is severe conflict between pedestrian and vehicular movement generally.

A plan recently prepared by the County Planning Officer includes proposals to divert through traffic, restrict the number of vehicular routes in the central area, provide rear service to shops where practicable and strictly control such service traffic where rear access would harm environment, or be unduly expensive.

Segregated pedestrian approaches are to be provided from the bus station, bus routes and car parks. The provision of car parks is to be kept in step with demand; if the projected demand for 3,100 spaces by 1985 is reached, multi-storey parks will be needed. These measures aim at conserving the attractive but

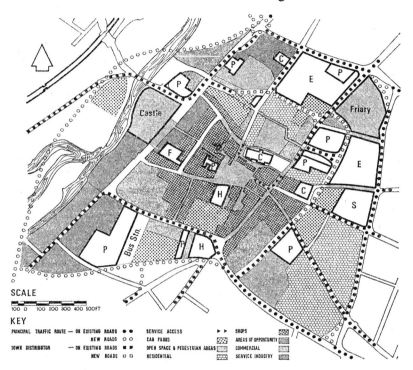

Figure 55. Newark: recommended revisions to draft town map.
C. Church: E. Education: F. Cinema: H. Hotel: P. Public Buildings: S. Swimming Pool.

lively character of the Market Place and shopping streets and the more subdued atmosphere generated by open spaces in the northern part of the central area, including those around the Castle, the council offices and the parish church (Fig. 55). On either side of Castle Gate 'areas of opportunity' have been defined. These contain buildings of high architectural merit which

223

it is proposed shall be restored and put to appropriate uses such as small shops, offices, restaurants, hotels or living accommodation.

Redevelopment in small towns where the population is likely to remain static is just as important as elsewhere; indeed in some respects it is more important if the town is to survive as a community. In the absence of the impetus provided by growth it is all too easy for a process of decay to set in which can only be reversed by strenuous effort. Such a town is Elland in the West Riding where for many years the centre had slowly been deteriorating. Congestion was caused by narrow streets, equally inadequate footpaths and by difficult intersections around Elland Bridge; this may be relieved if a bypass is constructed on the east side of the town. But the local council felt that it was impracticable to wait for decisions on regional road patterns. Redevelopment in the centre was necessary in any case and could be linked to any pattern providing the traditional connections with Elland's larger neighbours, Huddersfield and Halifax. The central area proposals prepared by Professor Denis Harper are for continuous reconstruction over a period of about ten years. The poor structural condition of most of the buildings justifies large-scale demolition; through traffic is mainly concentrated on the Huddersfield Road to the north and east of the centre. A new pattern could therefore be established, aiming at a higher quality than that which exists. Elland is not and can never be a regional centre; it could be a very attractive shopping centre on a more local scale. Distinctive character could be obtained by locating and designing new buildings to take advantage of interesting changes of level; this is evident from Professor Harper's proposals. Since there is little need for additional shopping, commercial and office accommodation, much of the central area can be used for houses and flats of different heights including two nine-storey 'star' blocks close to an existing recreation ground which would be extended, improved and replanted. The best of the old buildings would be retained and refurbished where necessary. Adequate vehicular access, parking and garaging could be provided for all parts of the scheme but pedestrian movement

48. Basildon New Town: shopping in the town centre with high block of flats in background.

49. Stevenage: a subsidiary shopping street.

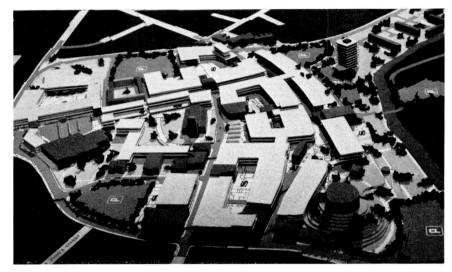

50. A model of the redevelopment scheme for Tamworth: Castle and river on right. Large-scale redevelopment is impracticable: pedestrian separation can be achieved without sweeping changes in the existing building pattern.

51. The start of urban renewal in the middle of Elland, Yorkshire.

would be given priority. This programme of renewal has been carefully phased; it is recognized that speedy rebuilding after demolition gives an important boost to morale and stimulates further effort. The attack on obsolescence, of which this is an example, is surely a sounder and more rewarding policy than that of outward sprawl, abandoning decayed central areas as a liability for a future generation (Plate 51).

The seaside resort, with a relatively small resident population but catering for a large influx of visitors during the summer, presents its own special problems. These become acute when one of the most attractive features of the town is its picturesque streets and groups of buildings. Such attractions are easily destroyed by brash, piecemeal rebuilding and efforts to provide vehicular access everywhere. A valuable example of how essential character can be conserved is the report on the Old Town, Hastings, prepared for the Borough Council by Lord Holford and R. A. Haskell. This is a street-by-street study of the Old Town which, in spite of its dilapidation and its inconvenience for modern traffic, retains great charm. Wholesale clearance would destroy its intimate scale and intricate pattern. Not many of the buildings are individually notable; groups and whole streets need to be considered as a whole in working out a policy of preservation, maintenance and sympathetic rebuilding. Detailed suggestions are made on this basis in the report. One of these, for the George Street/West Street area is shown in Fig. 56; it includes open and covered arcades of shops with living accommodation on upper floors. The scale is small, the character lively and reminiscent of the Lanes at Brighton. At the same time the car problem is not evaded; substantial car parks are proposed so as to eliminate the demand for parking on the beaches which would destroy Hastings' chief asset; service access and 'car squares' are allowed for behind shops.

The implementation of such a policy calls not only for further detailed study and design but for a co-ordinated and sustained effort over the years, employing all the varied powers of the local authority under planning, highway, housing and public health legislation; the council may also require financial help from the

Government on the recommendation of the Historic Buildings Council.

Another interesting project is that for North Berwick prepared by the East Lothian County Planning Department. The High Street of this small town is its main shopping street: it is also a main traffic route. Because of its narrow carriageway, coupled

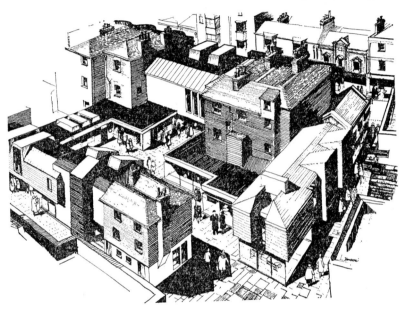

Figure 56. Architect's sketch of their suggestions for the George Street/West Street area, Hastings, comprising covered and open arcades connecting East Parade to George Street, shown here in the background with Market Passage to the right.

with footpaths only three or four feet wide, it has been a one way street since 1952. There is the usual congestion caused by the combination of through traffic, kerbside parking of commercial vehicles and private cars, and pedestrians, particularly in the summer. It is proposed that the High Street should be closed to traffic and that existing and new roads parallel to it be linked to form a circulatory system from which there would be service access to the back of the shops and to car parks. The character of the street would be maintained without impairing its commercial efficiency; to widen it would destroy this character, cost very much more and only slightly relieve the traffic congestion which will continue to grow (Fig. 57).

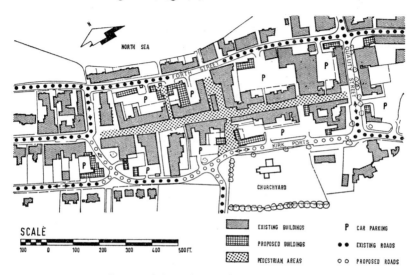

Figure 57. North Berwick: scheme for separating pedestrian and vehicular access to High Street.

These proposals have not yet been approved by the local authorities involved; indeed, they appear to have aroused considerable controversy. A scheme on these lines does, however, seem essential to allow another attractive and historic town to survive the impact of the car.

AN INTERIM STOCKTAKING

The complex problems of central area design are far from being resolved but some conclusions are beginning to emerge. The first is that it will be impossible to retain the traditional compact town centre and at the same time allow cars unrestricted access. Even

227

to provide for bus services, goods and service vehicles and other essential traffic, multi-lane roads and intricate intersections are needed. To cope in addition with all the private cars that people would like to bring in at peak hours these roads would have to be increased to enormous widths and the intersections bewilderingly complicated. The difficulties of accommodating moving vehicles are serious enough; as studies in Newcastle, Leicester, Liverpool and Manchester have shown, the parking problem would be even more intractable. The cost of unrestricted car access would be immense; at the same time public transport services would have to be maintained to carry those who do not own cars or do not want to use them for travelling to work: the resulting pattern would obviously be wildly uneconomic.

Whilst it is certain that many activities can more efficiently be carried on outside town centres and that improved means of communication by some form of television/telephone would encourage decentralization, traffic to the centre will continue to be very heavy. It can only be dealt with, on the one hand by restriction on entry to business and commercial zones and some form of toll on private cars, and on the other by a massive effort to improve community mass transport. Only in the smallest towns can we go on trying to have the best of both worlds by freely admitting all kinds of traffic.

Every town has special characteristics arising from its size, shape or geographical factors which will influence the choice of the most suitable form of mass transit. In some, trains—on the surface, underground or elevated monorail—will provide the answer; in others the emphasis will be on bus services; new types of vehicles which can run on rails or on the road are worth investigation. Buses for standing passengers only, mini taxis, moving pavements and escalators could all contribute to easier movement. Whilst cars are certainly the most convenient and in some ways the cheapest form of transport in areas which have low traffic volumes as in the suburbs, heavily-loaded routes can be efficiently served by mass transit systems designed for the purpose. But even cheap, fast and comfortable community transport will not be used unless services are frequent. This question has not yet received sufficient study. A wait of an hour for a long air flight is annoying but will be tolerated by travellers; a wait of a quarter of an hour for a five minute bus journey is much less acceptable and strongly encourages the use of private cars.

For shorter journeys in the town centre, legs are still the most flexible means of transport; moving pavements and escalators are obviously valuable aids to movement but can only be provided on the busiest routes. Here again the toleration factor comes into play: if people will only readily accept a walk of 150 yards from some form of transport to their place of work, the density of employment may be only 2,000 workers per acre in the centre of towns; if they will tolerate a walk of 250 yards it may be raised as high as 5,000 workers per acre.

Preoccupation with traffic problems and the cumbersome machinery for obtaining Ministerial approval of replanning schemes has very largely left the initiative in redevelopment in the hands of private investors. The criterion has therefore been profitability in its most restricted sense; new shops and offices have been built where and in the form they seemed most likely to produce a high return on capital. Few public buildings and amenities have been provided because they produce no observable revenue; it is in fact very difficult to measure the value of civic offices, concert halls and open spaces. Serious efforts are now being made to apply cost-benefit analysis to these essential elements, to the effect of one piece of development on another, and to the shift of site values. It is increasingly evident that no single planning proposal, whether public or private, can be regarded in isolation.[1] All are parts of a whole which must be co-ordinated: indeed what Professor Nathaniel Lichfield has called a 'planning balance sheet' is needed to assess the viability of capital investment projects.

It is evident that with so many inter-related factors at work, town centre design can no longer be left to the municipal engineer and the architect in a loose and somewhat uneasy partnership as

[1] Recognition of this by developers and traders is shown in the booklet *Shops in Redevelopment Areas* published by the Association of Land and Property Owners in 1966.

in the past. Town planners now need to collaborate fully with other specialists–sociologists, economists, valuers, traffic and service engineers and others, as local conditions require. This comprehensive effort cannot possibly be expected of any private organization, however well-intentioned; it is clearly one for the local planning authority if the public interest is to be served. The task is formidable enough: to make it more difficult still, every obstacle has been placed in the way in recent years. The repeal of the 'compensation and betterment' clauses of the 1947 Planning Act has caused the steep rise in the price of land, encouraged speculation in sites likely to be valuable and denied local authorities any substantial benefit from improvements which they have initiated. It remains to be seen if the new Land Commission will change this situation. There can be no real advance, no observable movement towards safer, more convenient and more beautiful urban centres until this situation is altered. The assembly of skill and the development of technique will be of no avail unless there are financial and administrative means of making them effective in practice.

The third important lesson to be learned from experience so far is the need to inform the general public about the purpose and processes or redevelopment. All such schemes inevitably entail acquisition of private property, hardship for some owners and tenants and at least temporary disturbance and inconvenience for many more. This is bound to arouse objections which can only be met by full explanation of schemes, adjustment where possible, and humane treatment of those who must be displaced. Where schemes have been fully illustrated and exhibited to the townspeople, and a flexible and human attitude displayed by the authorities, opposition has been minimized. It is necessary to go even further beyond this and engage public enthusiasm for, and pride in, the projects which are being evolved for their greater convenience and safety. This is an exacting task which calls for unremitting effort and patience but is a vital element in positive planning. The process of redevelopment is usually not one of dramatic wholesale clearance and renewal but of continuous change, carefully timed and programmed. A sympathetic climate of opinion cannot be hoped for unless it is constantly made clear how the pieces of jigsaw fit together. This need for better understanding between planning authorities and the public they serve is becoming increasingly recognized. In 1965 the Planning Advisory Group to the Ministry of Housing made recommendations for a new system of preparing planning schemes which involved a far higher degree of local responsibility and less detailed oversight by central government.[1] The Group pointed out in its report that: 'Some local planning authorities have already moved a long way from exclusive reliance on the statutory development plan as the medium for working out and publicizing their planning proposals. In some cases the need to translate their policies into the language of the conventional development plan only serves to obscure their intentions and delay the work. . . . Special care should be taken to explain not only the proposals but also the processes involved–the rights of objection, arrangements for re-housing, help with relocating businesses and other assistance that will be given to those affected. Public relations in this sense costs money but it must not be skimped and it is worth the expense of doing it well.'

[1] *The Future of Development Plans*, Report to the Minister of Housing and Local Government by the Planning Advisory Group, H.M.S.O., 1965. Many of the recommendations made in this report are incorporated in the Town and Country Planning Bill 1968.

CHAPTER 8

Towards Better Planning

In this final chapter an attempt must be made to draw together the threads of thought which we have reviewed and discussed. This involves some examination of the limits within which planning is practicable and the processes by which it can be effected.

National planning was introduced to Britain by the Romans: with their departure it lapsed and has only once been fully revived–in the Domesday Survey carried out in 1086 after the conquest of these islands by another relatively civilized race. Since then planning has been confined to very localized levels– the model villages of great landowners and industrialists, the Georgian and Regency streets, squares and crescents such as those of London, Edinburgh and Bath, and, more recently, plans for individual towns and cities. As has been shown earlier in this book the reaction against the social anarchy of the Industrial Revolution and the revolution in transport which accompanied it took the form of efforts to improve housing conditions which evolved into the garden city movement.

In 1913 a group of architects, engineers, surveyors and lawyers met to discuss the establishment of a professional body 'to advance the study of town planning and civic design, promote artistic and scientific development of towns and cities and secure the association of those engaged or interested in the practice of town planning.' This led to the formation of the Town Planning Institute in the following year: in view of the basic professional disciplines of the founders, their preoccupation with planning at a local level is hardly surprising. Whilst the doctrine of Patrick Geddes that survey and analysis of existing conditions was a pre-requisite to sound planning was accepted, his advocacy of

232

planning on a regional or national scale seemed too visionary to be practicable within the framework of British administration. Early regional plans such as those prepared by Abercrombie were advisory only and made little impact. A change took place in the 1940's: two separate streams of political thought and action joined. Concern with control of buildings and other forms of development had produced the Public Health Acts and early Planning Acts; the miseries of unemployment in the period between the two world wars had shown the need to provide new opportunities of employment in the areas worst affected such as South Wales and Tyneside. Local palliative action in each of these fields had failed. It had become evident that national and regional policies for industry were essential if opportunities for employment in the distressed areas were to be regenerated. Equally, better standards for housing, education, recreation and other social services could only be attained if considered in a regional rather than a parochial setting. This broader vision of physical and economic planning on a national scale is evident in the Barlow, Scott and Uthwatt Reports which formed the basis not only of the great Town and Country Planning Act of 1947 but of other legislation concerned with the distribution of industry and the establishment of new towns and national parks. Conditions seemed favourable for a great leap forward.

But during the 1950's the vision quickly faded. A number of new towns had been started: in spite of their success in the face of obstacles, no additional ones were designated. Some progress was made in establishing national parks; with one exception, they were left under the ineffective joint administration of several local authorities. The county councils, which had for the first time become planning authorities, prepared their County Development Plans, which were certainly a move towards a broader basis, but these plans were not co-ordinated with each other or with those for adjoining cities or county boroughs. The national policy was that the local planning authorities could work out their own salvation with general guidance from the various Ministries. Since the policies of the Ministries concerned with housing, employment, transport and agriculture were seldom co-ordinated—and sometimes seriously at variance—the traditional rivalry between local authorities, each striving for their own advantage, increasingly dominated the planning scene. The regional plans for Greater London, the West Midlands and other regions published at the end of the Second World War were denied the administrative structure which could have made them effective; in 1951 the regional planning offices of the Ministry of Housing and Local Government were closed. There was neither a national plan nor any positive move towards regional planning. This *laissez-faire* attitude could only have been justified in a state of equilibrium which did not exist. In fact, the population of the country as a whole has risen quickly in the last twenty years and is still rising: the distribution of population has been changing radically and is still changing. The growing use of electricity, gas and oil as sources of power frees many industries from concentrating around coalfields; as has been shown earlier, the increased use of road transport has widened the range of choice for the location of industry. The result has been a vast growth of population in the South-East of England and a lesser but still formidable increase in the Midlands, coupled with relatively slow growth or even decline in other regions.[1] At the same time, in all parts of the land, the car has encouraged the sprawling expansion of cities and conurbations and the use of the countryside for leisure activities on an unprecedented scale, whilst the rise in the use of motor vehicles for all purposes threatens the existence of public transport.

All this has been said over and over again by planners and other observers of the social scene for years; it is only very slowly being recognized that, as great humanists like Howard, Geddes, Corbusier and Mumford saw long ago, economic and physical development are inextricably linked and together produce the environment in which men live. Acceptance of the thesis that a comprehensive form of planning is practicable in a democratic

[1] For a more comprehensive account of present-day thought on regional planning in Britain see a series of articles by Donald Harris in *Official Architecture and Planning*, January to November, 1966.

society is even slower. Yet it is not only practicable: it exists already to some extent. Even in the United States, the citadel of faith in unplanned capitalism, the state plays a great part in controlling economic activity; it purchases over a fifth of the national produce; it employs vast technological resources in the development of space craft, supersonic transport, and warning systems for the detection of air attack; it restrains wages and prices and stabilizes purchasing power. By its nature, every modern industrial society, even where a capitalist ideology prevails, is increasingly compelled to manage the lives of the people whom it serves; whether or not we realize it–or even consciously oppose it –our beliefs and behaviour are guided by the needs and goals of the industrial matrix. The planned economy is with us already to a greater or lesser extent; the degree of democratic control over it varies in different countries and in different spheres of activity.

So far as housing policies and procedures are concerned there are encouraging signs that the extreme doctrinaire attitudes of capitalism and communism are receding. As has been pointed out by Professor Donnison,[1] whilst in Western Europe the mechanisms of the market remain important, governmental intervention, in the form of controls, subsidies and loans, deeply influences housing supply and demand, the structure of the market and its regional development. At the same time, in some of the communist countries of Eastern Europe, private saving and spending are increasingly encouraged to extend the range of choice in housing and relieve the state of part of the burden of housing investment. This convergence of policies and practice should prove interesting and make for fruitful interchange of ideas.

THE PROCESS OF PLANNING

In the first of his series of Reith Lectures in 1966[2] Professor J. K. Galbraith considered the effect of technology and its related requirements in time and capital on the modern economy, taking

[1] D. V. Donnison, *The Government of Housing*, Penguin Books, 1967.
[2] Published in *The Listener*, 17th November et seq., 1966. For a fuller treatment see also Professor Galbraith's *The New Industrial State*.

235

as an example the Ford Motor Company. The first Ford car came on to the market a few months after the company was formed in 1903: the company had an issued capital of £20,000 and 125 men were employed. In 1964 Fords introduced a new model, the Mustang; this had been in preparation for three and a half years; engineering and styling costs were over £3 million, tooling up had cost £18½ million, the average number of employees of the company was 317,000.

As Professor Galbraith pointed out, most of the effects of the increased use of technology are shown by these comparisons. An increasing span of time separates the beginning of the design of a car and its appearance on the market. Organized knowledge is brought to bear not on the manufacture of the car as a whole but on very small elements – on the qualities of particular steels or the machining of a particular part, then on the combination of these elements and so on to completion. 'The process of manufacture stretches back in time as the root system of a plant does down into the ground.' As a result of the increased time required, the increased investment in work in progress, the cost of specialized knowledge and its application to the development of complex and automated manufacturing processes, far more capital is committed to production. The first Ford required little of this: ordinary steels were used, no trained engineers were employed, all operations were manual. Now, 'time and capital tend to be committed more inflexibly to a particular task' which must be precisely defined before it can be divided into its component parts. If the task is changed, new knowledge and equipment has to be brought to bear on these components.

The sophisticated technology, planning and organization required to produce the new car each call for specialized manpower. This does not mean that supermen are needed: it does not mean that all members of the Ford organization must be infinitely versatile: it does mean that each must have a deeper knowledge of the specialized matters for which he is responsible. Co-ordination of the work of these specialists is itself a specialized task.

It follows, said Professor Galbraith, that this vast commitment of time and capital and the rigidity of these commitments calls for planning. 'Tasks must be so performed that they are right not for the present but for the point in time in the distant future when they are completed.' Modifications occurring in the interim must be anticipated so that their effects can be neutralized–or they must be prevented from happening. Unforeseen faults are extremely unpleasant and very costly.

The problems of national planning and the processes of putting it into effect are, of course, infinitely more complex than those of even a very large industrial company engaged in the manufacture of one single product with a limited life on the market. So far, a plan has been seen as a finite goal–a new shopping centre, a new road, even a new town of fixed size–to be attained in a period of x years at a cost of £y. In reality, planning is a continuous process: the goal to be reached is constantly changing so that it is more accurate to speak of a changing norm, changing with the evolution of ideas, skills and aspirations. Each intermediate 'goal' when reached is a factor which, fed back into the process of development, produces a new norm of civilization and influences the direction of the next move. The new shopping centre throws up fresh ideas for future shopping centres, the new road suggests improvement of the pattern of roads over a wide area, the new town provides experience for use in building more and better new towns. This process of continuous interaction between ideas and environment is not unlike the regenerative process employed by physicists in their pursuit of the absolute zero in temperature. A gas cooled through a spiral tube itself further cools the spiral containing other incoming gas, so that lower and lower temperatures are attained. In much the same way, improvements in housing, environment and transport generate demands for still higher standards: success in one field is a spur to greater effort in another. The present-day needs, ideas and aspirations of men crystallize as political policy. Before a policy can be efficiently implemented its implications in physical, sociological and economic terms must be examined and a plan evolved: the plan when put into effect produces a change in environment. And then the process begins again with new needs, new ideas, new aspirations.

There are, then, similarities and differences between the activities involved in producing a 'new' model of motor car and a new environment for people. Both require the application of organized knowledge to small elements and then to their combination into the completed product: both require specialized manpower: both require heavy commitments of time and money. The differences are, first, that at present the amount of time, manpower and money devoted to physical planning is very tiny in proportion to the task to be performed; nothing like the £21½ million spent over three and a half years in preparing for the marketing of the Mustang has been invested in research and analysis of environmental problems in the last twenty years, and what has been done in this field has not been co-ordinated. Second, and even more important, the physical planner's goal is not a product the design of which can be 'frozen', manufactured in large quantities and then discarded after, say, five years; it is a target which constantly changes itself from year to year; interim developments cannot be prevented and often cannot be adequately foreseen. Forecasts are therefore more fallible; some mistakes are inevitable and when made have to be lived with for a considerable time. Flexibility to reduce the ill effects of these in the long term must be combined with resolution and positive action to attain more immediate goals.

At all the stages of policy, planning and implementation, two different but closely interconnected judgements must be applied. The first is a judgement of balance and reality: can the desired result be achieved with the resources available? The second is an optimizing judgement, a judgement of value–is the solution offered the best answer to the problem set, having regard to all the standards of value which apply? Both are present in any situation in which a decision has to be made; in different situations either may dominate. Both types of judgement certainly apply to every planning problem.

A simple example is that of a borough council confronted with the common problem of traffic congestion in its town centre: there are several possible courses of action which may be stated in basic terms as:

Make-do-and-mend–More efficient, though restrictive, use of the existing street pattern by setting up one-way traffic systems, allowing parking for limited periods in minor streets, and providing temporary car parks on vacant sites.

Rehabilitation–Limited revision of the street pattern, closing of some streets to vehicles, constructing permanent car parks within a design which provides for redevelopment of selected limited areas.

Redevelopment–Wholesale acquisition and clearance of land in the centre, renewal on completely new lines and radical reconstruction of public transport systems.

Whilst future remedial measures are being considered the only immediate action possible is 'mend and make do': this is a simple reality judgement. Serious congestion exists or it would not be on the council's agenda: something must be done at once. So the no-parking and one-way street signs go up: any spare bit of land becomes a car park. But the council, if it is wise, will not leave the matter there; it will appreciate that palliative action will only be effective for a few years and will go on to examine the alternative long-term solutions. Is it practicable to adopt a policy of comprehensive redevelopment over, say, twenty years having regard to other expected demands on the council's resources within the same period? This again is a judgement of reality but if the answer to it is 'yes' judgements of value come into play. Should this central area redevelopment have priority over a by-pass road, slum clearance, improvements of the sewage works, provision of open space? Will redevelopment entail demolition of historic buildings or destruction of some other element of urban character?

There is, of course, a third choice: the council may value its centre so little that it decides to take no long-term action at all but to hope that somehow the problem will solve itself. This however is also a *decision*: inaction, as much as action, produces reactions in other fields of the council's responsibilities; it becomes one of the factors in the next problem which has to be faced.[1]

[1] An admirable study of the process of decision-making is *The Act of Judgement* by Sir Geoffrey Vickers, Chapman & Hall, 1965.

TOWARDS BETTER PLANNING

THE REGIONAL PATTERN OF BRITAIN

The last forty years has been a period of growth for the South-East of England – steady immigration from other regions in response to shortages of labour required in expanding industries, high earnings and a vast amount of building. Relatively, Scotland and the North have fared badly: there has been an outward drift of population, lower earnings, often serious unemployment, and little improvement of the urban fabric, particularly in the smaller industrial towns.

What has caused this shift of industrial balance is still far from clear and requires deeper study. The connection between this movement and other social processes is more apparent. A new factory in any given area means an immediate increase in employment opportunities leading to local scarcity of labour; wages will tend to rise attracting workers from less favoured parts of the country; the immigrants require housing, schools and shops, which create more jobs in the building industry. In the long-term there will be an increased demand for professional people, shop assistants, bus conductors and so on. The spiral process is further stimulated by the fact that migrants tend to be younger people with young families. All these factors build up a public image of buoyancy and progress.

Conversely, when a number of people become unemployed as a result of the closing of a factory they will often be faced with the alternatives of taking a job for which they are not skilled, and therefore less well paid, or moving to another region. It is the more enterprising and particularly the younger workers who will move, which not only deprives the community of its more lively members but reduces the rate of natural increase. The remaining population will tend to be relatively elderly and less receptive to new ideas, service employment will decline, the housing stock will not be renewed. The urban environment as a whole will deteriorate: the image will be that of a backward-looking community.

In Britain this imbalance cannot be accepted. Some of the natural resources on which the Industrial Revolution was founded will soon be exhausted; the greatest asset which remains is manpower. Unemployment and short-time working in some regions coupled with vacancies, unsuitable labour and high levels of overtime in others is appallingly wasteful. This is not to say that equality of growth between regions or within regions is desirable or even possible. Nor should it be assumed that the distribution pattern of industry of the nineteenth century should or can be restored.

The southward drift of industry and population and re-appearance of unemployment in less favoured regions in the early 1960's compelled the Government to take action. In 1963 two White Papers appeared outlining programmes for economic growth in North-East England and Central Scotland. Whilst these studies are somewhat sketchy they at least gave official recognition to an idea largely pioneered in Scotland – the 'growth area' as a corrective to national and regional unbalance. The theory is that if investment is concentrated on a selected area a process of growth is induced which later generates prosperity in the surrounding areas or region. Money earned in the growth area stimulates business in nearby towns, new firms become established and set up branch factories or foster the growth of linked industries. Thus a wide area, or even a whole region, is given fresh life and vigour. It follows that some other areas will not immediately benefit; some will never benefit at all but must face decline and decay; this aspect is harsh and more difficult to accept. The final tests are whether the national economy can afford to do more than establish limited growth areas and whether the climate of political opinion permits more to be done. Certainly no other positive theory has so far been put forward.

Since Patrick Geddes was a Scot, it is appropriate that Scot-land has led the way in exploring many aspects of national and regional planning. There are a variety of reasons for this: its legal system differs from that of the rest of Great Britain: the duties of the Secretary of State for Scotland embrace those of a dozen or so of his colleagues at Whitehall: the team of planners in his department survived when those of the other regional offices were broken up in 1951. So there have been greater oppor-tunities than elsewhere in official circles to study the problems of the country as a whole: whether one describes this as national or regional planning is a matter of taste—or nationality. These opportunities have been grasped, for Scottish planners have made distinctive contributions in the field of regional economic analysis especially with regard to the growth area theory. And the Clyde Valley Plan, published in 1949, was accepted as a logical way of alleviating extreme congestion of urban population, particularly in Glasgow, and led to the building of several of the earliest new towns.

The 'growth area' policy was taken a step further by a White Paper *The Scottish Economy 1965 to 1970* published in 1966. On the basis of the National Plan target of 25 per cent increase in national output by 1981 it is forecast that Scotland will have a shortage of people to fill the jobs available, even if the rate of emigration is reduced. New industry in the original growth areas has prospered; it is taken for granted that it will go on prospering and that the improved economy of Central Scotland will increasingly be self-generating. After 1970 it is proposed to extend the process to other parts of Scotland which are to be prepared for this in the meanwhile by 'holding operations' to arrest decline.

Whilst the growth area idea has penetrated into England and Wales it has not so wholeheartedly been applied. The Ministerial Studies of the South-East, the West Midlands and the North-West, published in 1964 and 1965, whilst useful in their recognition of the size and complexity of the problems of the regions, contained few firm proposals for their solution.

The South-East Study shows that the existing population of 18 million, south of a line running roughly from Peterborough through Banbury and Newbury to the coast, can be expected to rise by $3\frac{1}{2}$ million in the next eighteen years, of which $2\frac{1}{2}$ million will be by excess of births over deaths and one million by immigration from other parts of Britain and overseas. These increases will certainly take place unless present trends are checked or reversed. Indeed, the weakness of attempting regional planning without the framework of a national, physical and economic development plan could hardly be better shown than by the South-East Study. It is quite unrealistic to provide for continued large-scale increase of population in the South-East and also try to attract people to other parts of the country where population and employment are static or tending to decline.

The question of balance of population and employment between one region and another is not the only one needing decision; a policy for distribution *within* regions is equally essential. Even in this respect the South-East Study has proved to be inadequate. It assumed that the population of Greater London would remain static at about 8 million up to 1981. The outer metropolitan region, which extends to about forty miles from central London, and the rest of the South-East, including towns as far afield as Norwich, Northampton, Oxford, Swindon, Portsmouth and the coast resorts, would have to accommodate most of the natural increase of London as well as their own, together with some immigration from other regions and overseas. Much preparatory work has been done to get under way the government-planned expansion of Northampton, Bletchley, Southampton, Portsmouth, Ashford and other towns: it seems very unlikely, however, that these projects can solve the overspill problems of London in the next fifteen years. Only a small number of Londoners will move so far away: the function of these expanded towns will largely be to act as 'counter-magnets' to check further growth in London and to develop into sub-regional centres of more effective local government units.

The major contribution which is currently being made to new housing for Londoners takes the form of high flats, at excessive densities, in London itself. The supply of sites for these will soon come to an end: to bite into the green belt would be folly and could only aggravate the already intolerable congestion on roads and public transport, and the undue dominance of London as a centre. And more housing in *central London* is not needed: growth of employment in offices and factories has fallen off and the number of daily commuters has declined over the last few years.

It would seem logical to encourage the tendency towards dispersal of employment by working out a strategy of planned development in the outer metropolitan ring. Growth of population and employment is already taking place there on a large scale in a haphazard fashion. Without disrupting the green belt it would be practicable to concentrate growth in several areas in Essex, Surrey, Kent and Hertfordshire related to existing towns and communications. This would provide a pattern for the growth which will inevitably occur as a result of the pressures which exist or will arise: it would go far to solve London's housing problems and at the same time foster the correlation of employment and residence which has been successful in several of the new towns. Action on these lines is proposed in the first report[1] of the South Eastern Regional Economic Council—corridors of development towards the coast at Southend, the Medway and Southampton, and inland towards Peterborough, Northampton and Reading. Such action is certainly needed if complete chaos is to be avoided. The policy adopted until late in 1967 did not prevent the growth of population and employment in the South-East—and particularly outside central London—it only prevented it being planned.

Problems similar to those of South-East England arise in the West Midlands region where industry has gone on expanding. There is an unsatisfied demand for labour in spite of increases of workers through immigration from overseas and from continued migration from other parts of Britain. By contrast, the 'unfavoured regions' of Northern England, Scotland, Wales and Northern Ireland have much more acute economic difficulties expressing themselves in higher than average unemployment and a steady drain of manpower to the more prosperous South and Midlands. Between these two extremes are the 'grey' regions of Yorkshire and Lancashire where unemployment and migration are not so severe; consequently only parts of these regions may be areas where industrial firms are entitled to special incentive grants. But Yorkshire and Lancashire rely on employment in manufacturing and basic industries such as cotton and wool textiles, steel and coal, many of which are unlikely to require increasing numbers of workers in order to reach higher productivity. The faster-growing industries and those likely to need

[1] *A Strategy for the South East*, H.M.S.O., 1967.

52. The alternative to public transport: part of the freeway system around Los Angeles.

53. Transit expressway by Westinghouse. Cars on rubber tyred wheels couple automatically. Capacity is 20,000 passengers per hour.

54. Pedestrian conveyor up to elevated monorail station, Seattle.

larger labour forces are poorly represented in these regions, which increases their vulnerability in the long-term.

It was with the objective of giving increasing attention to regional development—ensuring the better utilization of the resources of all regions to the full by regulating the balance of investment between them—that the Regional Economic Planning Councils were set up from 1964 onwards in all the economic regions into which Britain has been subdivided. Their members were chosen by the Department of Economic Affairs from industry, trade unions, local government and the universities. It is too early to judge the effectiveness of these councils: what is certain is that they operate under severe handicaps. Their functions are advisory and in no way binding on the local authorities in their regions; they have little in the way of planning staff of their own; for professional advice they must rely mainly on the Regional Planning Board, consisting of civil servants representing various ministries. The boundaries of regions are far from satisfactory: because it was decided (with one exception) that existing counties should not be divided between two or more regions, towns like Sheffield and Bournemouth are separated from some of their important suburbs and commuter districts.

One by one the first reports of the Regional Councils are being published: only when all of them are available will it be possible to see how well they relate to each other. There is certainly far to go before it can be said that a National Plan exists which links economic and physical planning together.

PLANNING FOR TRANSPORT

In this context, the need for a national pattern of transport becomes evident. Here the Beeching Report, which aimed at 'making the railways pay' in isolation from other transport, was in conflict with the Buchanan Report which suggested the improvement of mass transport services, with the help of subsidy where necessary. Between 1951 and 1961, the number of people carried on the railways remained the same; travel by bus, expressed in passenger miles, fell by 15 per cent—in London by

245

more than a third and in other large towns by 19 per cent–and is still falling. If country and suburban railway services are to be withdrawn because they do not make a profit it seems likely that the bus services which are supposed to replace them will also be withdrawn or curtailed on the same grounds: this cannot make sense. There are areas where rail and bus services compete for passengers, particularly on long-distance services: these need careful scrutiny. But, where neither can 'pay', one or other must be provided as a public service. Without this, plans to revive industry in less prosperous areas must fail. Co-ordination of services is to be the main task of the Regional Transportation Boards which are being set up by the Ministry of Transport.

What of transport of goods? It would be foolish to deny that there are not many kinds of goods, products and materials which can go far more conveniently by road than in any other way. The important thing is to ensure that freight which could travel as easily or more easily by other means is not deterred by the gradual running-down or failure to develop other forms of transport. All possibilities must be used effectively: rail, road, air, water and pipeline. The development of coastal shipping might well be considered. The canals, in spite of a century of wilful neglect, carry a useful tonnage of bulk materials very cheaply and could carry more; the revival of selected lengths is far from being out of the question. The pipeline is a means of underground transportation with tremendous possibilities, not only for liquids or gas, but for solids which can travel in liquid. But the major contribution can certainly come from rail transport. Efficient marshalling yards, 'liner' trains and container services providing rapid and effective transference from road to rail and back again could immeasurably reduce the amount of heavy long-distance transport on the roads which has reached intolerable proportions.

A national road system is obviously essential: this is the age of the motor car and we must face the consequences. With a population slightly less than in Britain, and only 80 per cent of the vehicles we have on the road, West Germany has completed over 2,000 miles of motorways compared with Britain's 400 miles and has planned for 4,400 miles by 1970, compared with our 1,000 miles. Even allowing for efficient rail and water transport we probably need something of the order of 3,000 miles of motorway in the next twenty years linking the great conurbations, the industrial areas and the ports. But nearing the end of its journey, traffic will still need to come into towns and circulate round them: to deal with it comprehensive measures are required. Piecemeal operations such as Hyde Park Corner are no solution; they transfer the problem elsewhere. Growing preoccupation with traffic and mobility entails radical rethinking about the form of towns.

TOWNS FOR TODAY

The urban form which is most familiar is that of a central core surrounded by roughly concentric rings of development pierced by radial lines of communication, as in London, Paris and most other large cities, other than those on a sea coast. As the city grows, accessibility to the centre becomes more difficult. A variety of new forms are being tried out or studied. Howard's pattern of a central city with satellites has been carried out to some extent in the group of new towns around London. The linear plan of Soira y Mata is being developed far beyond even the Mars Plan for London:[1] the spine or communication corridor is looped back on itself or turned into a figure of eight as in the schemes for new towns in Buckinghamshire, in Northern Ireland and at Runcorn. These are all forms which substantially make use of one level only, though there may be multi-level town centres. Very sophisticated projects are now being worked out theoretically, notably in Japan, which use the third dimension of height and contain all the activities of a city in one gigantic structure within which complex arrangements of escalators, lifts, bridges and ramps link the various levels above and below ground.

Fresh thinking about the form of towns is certainly urgently called for. Not only is the population of Britain growing, but the increasing number of people will want more and better facilities

[1] See page 63.

for a greater range of activities, particularly with their longer periods of leisure: they already demand a greater degree of mobility. These different aspects of growth entail abandonment of the notion of a free-standing city or town divorced from its surroundings. The pyramidical pattern in which the central city or town serves the surrounding area has begun to break down already. Movement between home and other places is more varied, and more random than formerly: mobility provides choice over wider areas for employment, shopping, education and recreation and produces a web-like pattern completely different from the systems long established and still assumed to be valid. Most of these new forms indeed stem from a fresh attitude to traffic. It was formerly accepted that traffic routes were provided to serve land uses which existed or had been decided on. The tendency now is to adapt the pattern of land use to the needs of chosen modes of transport.

It is apparent that mass transport can carry mass movement to the centre of cities from the suburbs surrounding towns and countryside. Indeed this is how our great cities and towns grew up: they cannot go on without it, but they are being strangled by the private car—mainly in the form of car commuting with only one person per vehicle—even though this accounts for only a fraction of those travelling to and from their place of work. The prohibitive cost of providing for unlimited access by cars and—even more of parking them—is undisputed. Unlimited access cannot go on, whether we like it or not: as a nation we can no longer afford to allow the use of private cars in towns to be governed only by personal convenience at the expense of the inconvenience of others. We accept, though not always willingly, the restraint on an owner's use of land through planning legislation. We must now accept that restraints on the use of cars have become imperative on economic grounds alone.

It is, of course, futile to expect that people will cease struggling to get into town by car unless better alternative means are provided. Community transport will have to be made attractive. It must be convenient in terms of coverage, frequency and comfort; it must be as speedy as private transport—which today is not asking very much—and it must be relatively cheap. Much could be accomplished if the concept of transport 'making a profit' was dropped. The public must not expect to make a profit out of itself. The scales must if necessary be deliberately weighted in its favour by subsidy for the public good; we should not lightly discard the idea of community transport as a social service. Journeys to work, to central areas for shopping, for education and entertainment, must be mainly by some form of mass transport if we are to avoid complete chaos in the very near future. This almost certainly implies that new forms of mass transport will evolve. It is a good deal easier to invent new types of vehicles than new forms of towns. Monorail systems are being seriously considered in some of our cities, the hovercraft, which is now emerging from the experimental stage, may offer other possibilities if linked to a track. Mini-buses, new types of taxis and car hire—even provision of free bicycles for use in town centres as proposed by the angry young men of Amsterdam—are new but possible variations on more familiar themes to supplement mass transport.

Change in urban form is likely to be most apparent in the centre of towns and cities. Here, changes in shopping habits and cultural and recreational requirements are most clearly seen, with their repercussions in demands for land, on the location of buildings and the needs for transport. So far there has been a tendency to giantism in replanning. Larger shops, taller office blocks, more complicated road intersections occupying greater areas, have all been constructed, but the general pattern is the same. This cannot produce satisfactory results; the pattern itself requires reconsideration.

As has been said already the functions of central urban areas will vary according to the size and position of the city or town in its regional context. We need to examine more carefully the changing character of these functions and the effects of new mass media in the fields of entertainment and cultural activities which bring together people of large numbers. Theatres, concert halls, cinemas, and various kinds of sport have been affected by people's changing habits—probably most by television: new forms of entertainment such as bingo and ten-pin bowling have been introduced. Not only has the degree of support for such leisure activities to be reconsidered but also their possible shift of location within a town or city. Most industry and some businesses need not be near the centre at all and could move to smaller towns. In this regard, methods of communications are important; the telephone and teleprinter have brought great changes already. Further developments in communication techniques not yet apparent might reduce the need for commerce and business to congregate in a specific area and the need for personal visits to firms and individuals.

Flexibility in the planning of central areas, to take account of the rapid and continual process of social change and technological advance, is a prime necessity. To allow for this flexibility may mean that city and town centres will have to be planned with a more 'open texture'—the harking back to the very compact centre of medieval times, understandably beloved by architects for the wonderful townscape it produced, may be a mirage. Whatever the answers may be to the questions about urban centres, the concept of finality—of planning a centre for 'completion' by a certain date—must be abandoned. The centres of our cities and towns are in a constant state of change, the new replacing the old but itself also being replaced within a much shorter time than we have been used to expect. This is not to say that all the old will be swept away. In fact, the more open texture and the separation of pedestrians and vehicles in city and town centres should make it possible to keep much of our rich heritage of architecture and urban character. The first purpose of renewal should indeed be to recreate the idea of towns and cities as the foci of civilization: the centre must be a place in which people enjoy living, working and taking their leisure. This cannot be so long as town centres are regarded primarily as the happy hunting ground of speculators who are only concerned with profit. Probably much of the rebuilding must be done by private developers; at the same time it is ludicrous that profit should be the dominant factor. We deserve something better than this. In a very real sense the town centre belongs to its citizens: they have a right to expect to find there, in the harmony and contrast of building shapes and townscape, an expression of their corporate life as well as of their commercial activity, a touch of grace which takes no thought of ground rents, a vision as well as an investment.

RESIDENTIAL AREAS

The essential substructure of districts and 'neighbourhoods' in the older and larger towns and cities is difficult to ignore but some town planners have in recent years been suspicious about attempts to subdivide new communities. This is understandable: there was an interim period when the neighbourhood theory was applied rather unintelligently. Too large an area or too great a population was included in one unit: communal and social buildings which could have formed a focus were not provided. At a time when private transport makes it easier to choose between using local shops or those at a distance, to choose to work several miles from home and to maintain friendships and social contacts over a wide area, the 'failure' of the neighbourhood was not surprising.

Yet the neighbourhood idea constantly reappears under different names and varying forms: its persistence is perhaps the most striking demonstration of its validity. It appears, for instance, in the cellular pattern of towns proposed by Victor Gruen[1] and in the plans for the new towns in Buckinghamshire and at Runcorn and Washington which have already been referred to.

The search goes on for a smaller, more intimate social unit than the city or town itself. The primary school is still a factor in deciding the population of such a unit, although in the absence of other community facilities it seems difficult to justify this on purely social grounds. It may be that one should think in terms of units related to geographic and locational factors rather than school population; there is much to be said for providing some degree of overlapping between areas served by primary schools so as to allow a degree of choice of schools for parents. Certainly

[1] Victor Gruen, *The Heart of Our Cities*.

251

the size of neighbourhoods will vary with density and the degree to which they are purely residential. Rigidity of zoning for different uses–industry, offices, residential–is giving way to a greater flexibility. Some small industries and offices could be placed within the residential areas; not all 'neighbourhoods' will necessarily have the same texture. But the positive and valuable elements in the neighbourhood idea are worthy of further detailed study and experiment. There is still, for instance, almost complete failure to give the smaller community units any democratic political significance–a failure which is perhaps symptomatic of a debased level of local democracy rather than of the technical difficulties. A small district which has social coherence and recognizable boundaries is surely a sound basis for an electoral ward.

It is of prime importance that the design of residential areas should be studied intensively: nearly half of the houses in Britain are over fifty years old and it has been estimated that nearly fifteen million new ones–nearly equivalent to all our existing stock–will be needed by the end of the century. There are still vast numbers of slums–houses that have been officially declared to be unfit for habitation–and many more that should be condemned because they are worn out. Most of the houses in this country are inconvenient and poorly equipped; only a very few of them are properly heated; renewal and improvement have not kept pace with the rate of decay and obsolescence. Nor have there been any great advances, on a national scale, in design and layout. Many speculative builders have blown the dust from their drawings of the 1930's and re-used them. The more alert local councils have done better and have experimented both in house design and in layout; the less enlightened have gone on reproducing the pre-war pattern in a tighter form.

The new towns have shown the real nature of the problem. Here, large numbers of houses have been erected using traditional building methods under relatively ideal conditions, unfettered by existing development. The designs and layouts, though much better than the general standard of local authorities or private developers, show that the widespread ownership and use of

55, 56. Bishops Stortford: High Street and Bridge Street
as existing and as proposed.

57. Shopping in safety in the central precinct at Coventry.

58. Parking on flat roofs of shops and market at Coventry.

motor vehicles, which overtook the new towns in the course of their construction, call for fundamental changes. The old corridor-street pattern cannot be effectively adapted to the new circumstances. Garaging for every family, parking space for visitors, access for tradesmen–all making their demands on space, increasing the hazards of movement and invading domestic peace and privacy, require a completely fresh type of layout. The logic of the Radburn principle, providing a system of footpaths from homes to schools, shops and open spaces and underpassing all traffic routes, is now accepted in theory. Yet the places where it has been fully implemented are comparatively rare. The new towns that have experimented with pedestrian and vehicle separation have clearly demonstrated that to apply these principles to a small portion of a residential area is merely to tinker with the problem: the separation of vehicle and pedestrian should be applied on a very large scale–if possible to the whole town, as has been done at Cumbernauld and is being done in other new towns. At the other end of the scale, the type of layout now required entails changes in the design of the dwelling itself: as was shown in earlier chapters, it is not satisfactory to seek to use old housing types and adapt them to these new forms of layout. The two–the dwelling internally and its external environment–must be considered together. Higher standards of space, privacy, convenience, fittings and equipment are required: the recommendations made by the Parker Morris Committee in 1963 are only slowly being put into practice.[1]

These changing standards lead one to consider whether we do not build too permanently. In the U.S.A., it has been said, the economic life of buildings is as short as thirty years. In Britain it is certainly longer than that but tending to shorten. There is no move yet to pull down and replace houses built in the 1920's and 1930's but this may very well come in the foreseeable future.

It is becoming recognized that the vast number of houses that

[1] Statistics issued by the Ministry of Housing and Local Government in 1967 show that only about half of the houses erected by local authorities complied with the minimum standards recommended.

are required can only be provided by industrialized, mass production methods. This is bound to bring with it standardization either of large elements or of complete house designs, which not unnaturally arouses fears of barrack-like monotony of appearance. Here we may ask ourselves if there is really any individuality and variety in the average housing estate with its corridor streets lined with houses on a standard building line and different only in small details and the colour of paintwork. Much greater variety and individuality could be attained, even with completely standard house designs, if the layout and siting of the houses were more carefully studied and related to landscape, levels and planting; these are the keys to humane living conditions.

The debate on residential density goes on, usually on an emotional rather than a rational basis. It is generally agreed that families with children desire, and need, a garden–but not necessarily a large garden: many childless families, and single people who want their own homes would be pleased not to have one. These varied requirements can be met, given a willingness to consider and experiment with types of houses and layouts which are not always on accustomed lines.

PATTERNS OF CHANGE

At the moment Britain is not enjoying to the full the benefits of industrialization and the rich potential of urban living. The aftermath of the Industrial Revolution, followed by the Transport Revolution and two world wars, has tried our resilience. These great upheavals, when the apparently impossible happened, have been succeeded by a period in which everything seems too difficult to be tackled. Evasion of reality has infected most aspects of life. Industry and commerce could benefit from the employment of scientists, technologists and, indeed, trained minds and hands of every kind, but the growth of facilities for training is all too slow. There has been a flow of skilled people to richer pastures abroad and dissatisfaction among those who remain because of the inadequate use of their abilities and the miserly allocation of public and private money for research and development. Restrictive practices abound in all walks of life. Attachment to out-of-date machinery and methods infects not only industry but administration and government. The tangled web of Government departments and *ad hoc* boards holds together mainly because of the genius of the Civil Service for making any system work, matched only by its capacity for pointing out the difficulties of making any change. Our totally outmoded and inadequate system of local government, at its best, struggles to cope locally with social problems which are, in reality, national issues. These problems are indeed formidable—15 million houses to be built, four times as many vehicles on the road, a population increase of 20 million in a country already overcongested in some regions, and a distribution of population and industry which is economically and socially out of balance. Solution of these problems would tax the resources of a highly efficient administrative system—and our administrative system is far from efficient.

Redistribution of population and industry is a task which can only be performed by the national Government: the production of more cars can, with complete confidence, be left to the manufacturers. This still leaves a vast field of administrative and productive activity much of which, in theory, is within the ambit of local government.

It is commonplace that local government has been in decline for nearly forty years. Many of its functions have been taken away and vested in *ad hoc* regional boards—electricity and gas supplies, hospitals and in many cases public transport. Public assistance, and valuation and assessment for rating, are now in the hands of Government departments. More than half of the money local councils have to spend comes direct from the national exchequer: because of this practically all expenditure and every decision has to be referred to Whitehall for approval. As one town clerk has put it, councils have freedom of action only in the collection of rates and refuse. The exercise of initiative and power has steadily been centralized.

Local government is equally hampered by its complex

structure of counties, county boroughs, municipal boroughs, urban district and rural district councils. This leads to an unreal conflict of interest by cutting off, administratively, towns and cities from the areas surrounding them. Recognition that the conflict *is* unreal and that many local government activities can only be effectively tackled in a broader basis has brought into existence consortia of local authorities for the design and erection of schools and houses: in this way organizations have been formed which are large enough to attract adequate qualified staff and take advantage of industrialized methods of building on mass production lines. These *ad hoc* bodies are useful expedients to make 'local' government go on functioning somehow but an end to such tinkering is long overdue. The whole concept needs to be changed – and this change should be not by further centralization of responsibility but by devolution from Whitehall to a regional level.

It is not within the scope of this book to discuss in detail the question of local government reorganization but we feel bound to raise the fundamental issue of democratic control over planning. We have tried to show that planning is essential to the development of our society. As the poet W. H. Auden has written: 'Virtues which were once nursed unconsciously by the forces of nature must now be recovered and fostered by a deliberate effort of the will and the intelligence. In the future, societies will not grow of themselves. They will either be made consciously or decay. A democracy in which each citizen is as fully conscious and capable of making a rational choice, as in the past has been possible only for the wealthier few, is the only kind of society which in the future is likely to survive for long.'

This presents a dilemma: it is more and more difficult for the ordinary citizen to make a rational choice on many problems which are vital to his welfare: he must rely to an increasing extent on skilled advice from specialists. In theory, at the moment, the elected members decide on policy and the officers implement it in detail: in practice, members, particularly of small authorities, engross themselves in detail and tend to let policy evolve from precedents formed by detailed decisions over a period. When an important policy decision has to be made it may be taken without full consideration or knowledge: it is more likely that the prohibitive procrastinator, identified by Professor C. Northcote Parkinson, will have his way. His technique of negation by delay is never to say 'yes' or 'no' but to propose the appointment of a committee to consider the need for a policy; this committee sets up subcommittees to consult other bodies and report back on various aspects of the problem. If it is decided that a policy is required—and this is by no means certain—another committee is appointed to advise what it should be. This game is capable of infinite refinement and convolution: played with sufficient skill it can even involve an approach to a Minister of the Crown who may attempt, with greater or lesser enthusiasm, to get a reform incorporated in his party's programme. Eventually, in an emasculated form, it may pass into law, usually long after its originators are dead. A classic example, though not connected with local government, is the almost imperceptible movement towards the adoption of decimal coinage proposed by Gladstone in 1853, which is at last to be put into effect—more than 100 years later.

Not all these delays arise from deliberate obstructionism: as often as not they stem from a laudable desire to take into account all points of view. Indeed, at every level, the corridors of power, like the road to hell, are paved with good intentions. These are not enough: as the interrelation of all aspects of national and local government becomes clearer and the field of their activities grows, specialist knowledge becomes more essential and lay opinions less valid as a basis for policy decisions. As with the Ford Motor Company, quoted earlier, skilled minds are required to analyse small elements of the total problem, to assemble these detailed findings and produce as complete an answer as possible. The design of the Mustang was not decided by the board of directors: the quality of steel to be used and the speed of the assembly line were not the subject of votes at a general meeting. All these and an infinite number of other matters were decided by specialists employed for their ability to arrive at informed decisions and invested with the power to put them into effect. In effect, the directors, and, in theory, the shareholders, made only two decisions—first, that a new model of car be produced and second, having seen the prototype, that it should be put on the market at a certain price. The evolution of a similar pattern in local government may be delayed but is inevitable. It is easier to design and make cars than to build and administer cities: if the layman is only able to exercise very general control over a company which he partly owns it is evident that he will be much less able to decide how to get the best out of urban life. Much more discretion and freedom must be entrusted to salaried professionals to decide and to act: it follows that they would bear a heavier responsibility for failure and indecision. All the same the elected councillor has an essential role, both as a democratic watchdog and as a two-way channel of information between electorate and officials. The ivory tower attitude of the nationalized boards shows what happens when this vital link is missing.

Applying this line of thought to the subject of this book we do not claim that planners are omniscient; town planning techniques are still very imperfect: in an historic sense they are only just emerging. Application of the first modern theories on the conscious design of towns has been overtaken by the more rapid evolution of the motor vehicles and, in particular, by the wide use of the private car. The work of such pioneers as Geddes, Howard, Unwin, Perry, Abercrombie and Le Corbusier is only now beginning to bear fruit. And the problems to be faced are not static: they change and grow more complex from year to year. Nevertheless they are no longer beyond man's comprehension, given the essential tools. The first of these is facts: too often planners have to work on inadequate information. Despite all that can be said against statistics, they are the most reliable basis for the measurement and relationship of facts on which judgements of reality can be made. In Britain vast quantities of figures are available on every subject but too often it is incredibly difficult to relate them to each other. Population census and housing figures relate to local government areas which do not coincide with the areas served by Ministry of Labour employment exchanges: Board of Trade figures for industry are not always easily related to either. Close co-ordination in these respects would be immensely helpful and eliminate the need to make estimates which can be highly erroneous. Given reliable figures, they can now be processed by means of electronic computers in a way which a few years ago would have been impossibly laborious: relationships between groups of statistics appear in a very short time.

Another essential tool is a fuller understanding of value judgements. Planning is for people and we need to know very much more not only about how they live and work and move but how they would like to live and what they most value in urban life. These also are facts, though much more difficult to ascertain. To get anything like a rational answer it is necessary for people not only to see what exists but what might be if they want it and what the alternatives are in fulfilling their desires. Here again computers can be of great assistance in working out the effects of different possible patterns for industry, road and rail networks and social services, and in formulating policies about population distribution and the size of towns. This does not imply a pattern imposed by computers: their value is in calculating the effects of alternative courses of action and allowing for the desire for diversity. Even in those respects computers have their limitations: whilst they can answer many questions, man alone can decide what questions to ask and how and when to ask them. The collection and processing of figures is not a substitute for thinking.

This fuller information, these more adequate tools, would not alone produce the answers: they would greatly help in exploring acceptable ways towards a more humane environment. Most would, however, be learned from experience gained in actually building new towns and cities: all calculations and research are invalid if they do not stand up to practical test.

Further progress is hampered not so much by lack of ideas or resources as by habits of mind and processes of administration. The fate of the Buchanan Report is an instructive example. One of its most notable characteristics was, paradoxically enough, that it did not pretend to offer a cure-all for traffic in towns. The Report analysed the problem on the basis of ascertainable facts and their relationship to each other and set down a series of judgements of reality: it identified three variables–accessibility, environment and cost: it worked out the implications of applying assumed standards of accessibility to existing situations, using a variety of innovations such as raising buildings above ground level so that vehicles could move more freely below. But it was not suggested that a perfect answer, applicable to all situations, had been discovered and could be recommended. Instead it was left to the Government and the public to choose: if a very high degree of accessibility for cars in all or some parts of a town was demanded, the cost, financially and in environmental quality, would be correspondingly high; if environment was greatly valued, car traffic would have to be severely limited; the Buchanan Working Group did not set itself up to the community as an arbiter of values. Here at once lay the strength and the weakness of their report: its virtue was the cool and objective way in which it pointed out various possible courses of action, its weakness was its lack of power to propel public opinion towards making a choice. In the absence of any authority capable of choosing and putting its decisions into operation, the Buchanan Report, everywhere acclaimed and 'accepted', has been pigeonholed. Its steering body, the Crowther Committee, recognized the lack of such an authority and tentatively suggested what form it might take. In fact, of course, the only really effective body would be an all-purpose regional authority.

The inevitability of change–and increased rate of change–is now an accepted fact. 'The price of liberty is constant vigilance': the price of good environment is no less. Everything which militates against it must be challenged. We must even question the continued existence of the motor vehicle as we know it. Fifty years ago there were few cars: fifty years hence it is certain that they will have changed out of all recognition or even been superseded by another form of transport. Yet economically we depend heavily on the car; production and home sales increase; the export of cars forms a high proportion of our trade. Vehicles today play the part in the economy that cotton did at the end of the last century but, as the decline of the cotton trade showed and fluctuations in car exports remind us, Britain cannot indefinitely rely on one product which can be made elsewhere. Export of 'quality' cars and lorries may continue but sooner or later every major country will mass-produce its own cars; the home market will be almost the only market. Is this thinking too far ahead? Can we afford even now to assume that 'the sky is the limit' in car manufacture and distort our national economy accordingly? These are questions which town planners cannot be expected to answer but are entitled to ask, since they are at the receiving end of the process of economic and technological development and are vitally concerned with its effects.

The conflict grows between the needs of urban man for good environment – peace and privacy at home, space, vitality and safety in public places – on the one hand, and mobility on the other. It has so far mainly been apparent in the West – in Europe and the United States – but has now arisen in other parts of the world – notably the highly industrialized cities of Australia and Japan – and can be expected to extend further. In countries with a high degree of control over economic planning, such as the U.S.S.R., and the countries of Eastern and Central Europe, control over production of motor vehicles may make it possible to avoid a crisis. But imitations of the West, in efforts to equate economic parity with parity in status, could lead to similar errors being made. Certainly in the Western World the motor car, a symbol of affluence and mastery over distance and time, has become a dictating force, making rapacious demands upon human energy, resources and space. Man has become the 'sorcerer's apprentice', struggling desperately to keep pace with the monster he has created.

Recovery of a sense of order in human environment is essential – and it is attainable, given the will to attain it. There is no lack of ideas, many of them well-tried already but not widely applied. There is no lack of resources if expenditure on non-essentials or items of lower priority is any guide – expenditure on space travel, supersonic aircraft, colour television, bowling alleys, bingo and betting. Undoubtedly it will be costly, not only in financial terms but in discipline of established habits, to resolve

261

the conflict between environment and mobility. But we pay now, very heavily indeed, for a highly inefficient society: we pay in deaths and injuries, anxiety and tension, which cannot be financially valued, as well as in land prices, taxes, fares, freight charges and delays. We get a very poor bargain: at no greater cost–and possibly less–we could reshape an environment which would be efficient and exciting to live in, the expression of a confident belief that people matter most. It is a question of values: in the words of the old Spanish proverb: 'Take what you want', said God, 'take it–and pay for it.'

262

Further Reading

Some of our readers will, we hope, wish to go further into one or other of the themes we have discussed. A great deal that is relevant has been written in the last decade–in other books, articles, Government publications, reports and academic papers. A number of these have been quoted or referred to and only a few of them need be mentioned again here. This note suggests others which are likely to be the most useful to laymen, councillors or professional people trying to keep abreast with current thought in this increasingly complex field and to students seeking further introduction to it. The books chosen are all in print, most of them in editions published in the last few years: all should be obtainable from bookshops or in university or public libraries.

An understanding of social history is essential to any sound appreciation of the aims of town and regional planning. There are many excellent social histories of our own century: one of the best for this purpose is Pauline Gregg's *Social and Economic History of Britain* (Harrap; revised edition 1966). By far the most wide-ranging study of urban development in Western Europe and the U.S.A. is still *The City in History* by Lewis Mumford (Secker and Warburg and Harcourt, Brace and World, 1961). *Town and Country Planning* by the late Sir Patrick Abercrombie, third edition, revised by D. Rigby Childs (Oxford University Press, 1959) provides useful background. The current problem of population pressure and the resulting growth of conurbations is concisely and urgently treated by Peter Self in *Cities in Flood* (Faber, 1961; Transatlantic Arts).

Those who wish to follow the development of garden city ideas may do so in the current edition of Howard's *Garden Cities of To-morrow* and the introduction to it by Lewis Mumford (Faber 1946; Transatlantic Arts, 1951). Much of the work and thought of Le Corbusier is covered in *The Radiant City* (Faber and Grossman, 1967).

An earlier book on the evolution of the problems of the motor vehicle in Britain is *Mixed Blessing: The Motor Car in Britain* by C. D. Buchanan (Leonard Hill, 1958). This was a pioneering study which led to the major study, now universally recognized, *Traffic in Towns*, prepared by a team directed by Professor Buchanan (Ministry of Transport, H.M.S.O. 1964), which deals with the nature of the problem of urban traffic, establishes a working theory and relates this to the urban situation in British towns and cities of different sizes. Statistics of vehicles, traffic, roads, accidents, etc., are collected together annually in a very valuable little publication, *Basic Road Statistics* (British Road Federation, annually). For those who are undeterred by crusading zeal, Paul Ritter's *Planning for Man and Motor* (Pergamon Press, 1963) contains a great deal of information on the conflict between vehicles and pedestrians, and illustrates attempted solutions all over the world. A lively and comprehensive account of modern and possible future modes of transportation is *New Movement in Cities* by Brian Richards (Studio Vista and Reinhold, 1966).

The social and community aspects of new towns are now receiving increasing attention. An earlier book deserving notice is derived from a special study; this is J. H. Nicholson's *New Communities in Britain* (National Council of Social Service, 1961). The application of the lessons of building the first fourteen new towns and a detailed and well-illustrated study of a new town for Hook in Hampshire is the subject of *The Planning of a New Town* (London County Council, 1961). *Economic Planning and Town Expansion* by J. H. Dunning (Workers' Educational Association, 1963) is a special study of the problems of industry and employment for large-scale town expansion based on a case study of Basingstoke, Hampshire. Two studies of new towns that have been published in book form are recommended for the clarity with which the proposals are presented and for the high quality of illustrations and the careful explanations of the principles underlying the proposals. These are Arthur Ling's

FURTHER READING

Runcorn New Town–Master Plan (Runcorn Development Corporation, 1967), and *Washington New Town Master Plan and Report* by Llewelyn-Davies, Weeks and Partners (Washington Development Corporation, 1966). A study of the regional economic and social problems which should provide a context for a new town is provided by the first volume of *The Lothians Regional Survey and Plan*, concerning Livingstone New Town in central Scotland, by a team of contributors from the University of Glasgow and edited by D. J. Robertson (H.M.S.O., 1966). While some of the material contained is of a specialist nature and form of presentation (as is the second volume dealing with physical planning aspects) some of the chapters and especially the introduction and concluding report on *A Policy for the Lothians*, by Professor Robertson, illustrates a broader approach to the relationships of new towns to regional planning.

Regional studies are now being issued for each of the economic planning regions of Britain. These outline the economy of each region and analyse the specific population, employment, housing, communications and environmental problems of each region as the raw material required to help formulate policies and provide solutions.

Peter Hall's *The World Cities* (Weidenfeld and Nicolson, World University Library and McGraw-Hill, 1966) is an excellent description of the metropolitan planning problems of seven of the great 'city regions' of the world and a discussion of planning policies and their relative success, whilst his *London 2000* (Faber, 1964) deals with the complex problems of urban expansion in London and their possible solution.

Economic and political aspects of housing and urban renewal are dealt with more fully in several books. Foremost amongst these is D. V. Donnison's *The Government of Housing* (Penguin Books, 1967). For an economic analysis of the housing situation, *The Economics of Housing* by Lionel Needleman is suggested (Staples Press, 1965). Two books by J. B. Cullingworth describe very clearly the present form of town planning administration, *Town and Country Planning in England and Wales* (Allen and Unwin, 1964; University of Toronto Press, 1965), and the work of local authorities in the housing field and the way housing policy has developed, especially since 1945 in *Housing and Local Government in England and Wales* (Allen and Unwin and Humanities Press, 1966). A shorter description of the system of town and country planning in Britain is the primer by D. W. Riley *The Citizen's Guide to Town and Country Planning* (Town and Country Planning Association, 1966).

On the design aspects of housing and housing layout, the Parker Morris Report *Homes for Today and Tomorrow* (H.M.S.O., 1961) is still a valuable source of reference as also are the Design Bulletins and the Planning Bulletins produced from time to time by the Ministry of Housing and Local Government. The study of the problem of renewal in outworn areas, where older houses need refurbishing and the environment improving, was the subject of a special study at Rochdale by the Ministry of Housing and Local Government *The Deeplish Study* (H.M.S.O., 1966).

Landscaping around buildings is admirably treated in Elizabeth Beasley's *Design and Detail of the Space between Buildings* (Architectural Press) and *Landscaping for Flats* (Design Bulletin No. 5, H.M.S.O., 1963), whilst the visual aspects of civic design in its broadest terms can be studied in *Townscape* by Gordon Cullen (Architectural Press, 1961; Reinhold, 1962).

Sources of Photographs

Acknowledgement is made to the public authorities, architects, agencies and journals who have kindly allowed us to use the photographs indicated. The name of the photographer, where known, is also acknowledged in brackets.

Plate

1.
2. } Instituto Italiano di Cultura
3. John Tetlow & Partners
4. National Monuments Record
5. Scottish Tourist Board
6. British Road Federation
7. P. A. Reuter Photos Ltd.
8.
9. } United States Information Service
10. Harlow Development Corporation (Wainwright)
11. Harlow Development Corporation (John McCann)
12. Cumbernauld Development Corporation (Bryan and Shear)
13. Basildon Development Corporation (Henk Snoek)
14.
15. } Cumbernauld Development Corporation
16. Clifford Culpin and Partners
17. Basildon Development Corporation (Henk Snoek)
18. *Architect's Journal*
19. London County Council and Norah R. Glover
20.
21. } Harlow Development Corporation (John McCann)
22. Skelmersdale Development Corporation (Arthur Winter)
23. *The Guardian*
24. Aerofilms and *Aero Pictorial*
25. London County Council

Plate

26. Ministry of Housing and Local Government
27. Coventry Corporation (P. W. and L. Thompson)
28. *Architectural Review* (William Toomey)
29. City Architect, Sheffield (A. N. Tonley)
30. John Madin and Partners (Reilly and Constantine)
31. Eric Lyons (Peter Pitt)
32. Eric Lyons (Henk Snoek)
33. Greater London Council
34. Chamberlin, Powell and Bon (Alfred Cracknell)
35. City Architect, Sheffield (Roger Mayne)
36. City Architect, Sheffield (J. A. Coulthard)
37. Chamberlin, Powell and Bon (Alfred Cracknell)
38. Coventry Corporation (P. W. and L. Thompson)
39. Netherlands Embassy (Doeser Fotos)
40. ⎫
41. ⎭ Royal Danish Ministry for Foreign Affairs
42. ⎫
43. ⎭ City Planning Officer, Newcastle on Tyne
44. ⎫
45. ⎭ Graeme Shankland Associates (John Mills)
46. City Planning Officer, Leicester
47. Coventry Corporation (Vivian Levett)
48. Basildon Development Corporation
49. Stevenage Development Corporation (G. L. Blake)
50. John Tetlow & Partners (J. Garfield Snow)
51. Abbey and Hanson, Rowe and Partners
52. United States Information Service
53. Westinghouse Electric International Co.
54. Central Association of Seattle and Brian Richards
55. ⎫
56. ⎭ County Planning Officer, Hertfordshire
57. ⎫
58. ⎭ Coventry Corporation (Vivian Levett)

Index

INDEX

INDEX